Improving the Sensory, Nutritional and Technological Profile of Conventional and Gluten-Free Pasta and Bakery Products

Improving the Sensory, Nutritional and Technological Profile of Conventional and Gluten-Free Pasta and Bakery Products

Editor

Barbara Simonato

MDPI • Basel • Beijing • Wuhan • Barcelona • Belgrade • Manchester • Tokyo • Cluj • Tianjin

Editor
Barbara Simonato
Department of Biotechnology
University of Verona
Verona
Italy

Editorial Office
MDPI
St. Alban-Anlage 66
4052 Basel, Switzerland

This is a reprint of articles from the Special Issue published online in the open access journal *Foods* (ISSN 2304-8158) (available at: www.mdpi.com/journal/foods/special_issues/Gluten_free_Pasta_Bakery_Products).

For citation purposes, cite each article independently as indicated on the article page online and as indicated below:

LastName, A.A.; LastName, B.B.; LastName, C.C. Article Title. *Journal Name* **Year**, *Volume Number*, Page Range.

ISBN 978-3-0365-1290-7 (Hbk)
ISBN 978-3-0365-1289-1 (PDF)

Contents

foods

MDPI

Editorial

Improving the Sensory, Nutritional and Technological Profile of Conventional and Gluten-Free Pasta and Bakery Products

Barbara Simonato

Department of Biotechnology, University of Verona, Strada Le Grazie 15, 37134 Verona, Italy;
barbara.simonato@univr.it

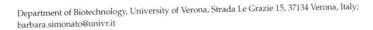

check for updates

Citation: Simonato, B. Improving the Sensory, Nutritional and Technological Profile of Conventional and Gluten-Free Pasta and Bakery Products. *Foods* **2021**, *10*, 975. https://doi.org/10.3390/foods 10050975

Received: 25 April 2021
Accepted: 26 April 2021
Published: 29 April 2021

Publisher's Note: MDPI stays neutral with regard to jurisdictional claims in published maps and institutional affiliations.

There is currently a growing consumer interest in healthy food. Cereal-based products such as pasta and baked goods represent staple foods for human nutrition. Due to their worldwide diffusion, these products can be carriers of nutrients and bioactive compounds; therefore, they lend themselves very well to the fortification process. Furthermore, among new formulations of cereal-based food, gluten-free products have become popular even among people without celiac disease, who have chosen a gluten-free lifestyle. The improvement of well-being, sustainable lifestyles, and waste control are also aims of the United Nations for the Agenda 2030 (UN 2015), which has motivated food scientists and industrial producers to research new and healthier formulations for pasta and baked goods preparation. In this context, researchers are also encouraged to use agro-industrial by-products of high added value for food fortification. In this frame, the Special Issue "Improving the Sensory, Nutritional and Technological Profile of Conventional and Gluten-Free Pasta and Bakery Products" collected nine original articles focused on new product formulations of gluten-free pasta or baked products formulation, as well as agro-industrial by-product utilization. The final aim was the preparation of valuable products from a nutritional, technological, and sensory standpoint.

Cappa et al. [1] studied the effects of red rice or buckwheat flour inclusion in potato-based pasta (gnocchi) concerning the nutritional, technological, and sensory characteristics of the final product. The researchers concluded that gnocchi fortified with buckwheat flour showed better texturizing characteristics and could benefit from the claim "source of fibre". Nevertheless, wholemeal buckwheat is responsible, due to its particle sizing, for a negative response from a sensory point of view, thus reducing the acceptability of the fortified gnocchi.

As part of the production of healthy foods, Krupa-Kozak et al. [2] produced a gluten-free sponge cake by replacing sucrose with fructans of different degrees of polymerization, to observe its technological and sensory characteristics. The authors demonstrated that sucrose is not necessary to obtain a gluten-free sponge cake with favorable technological and sensory features, as the one prepared with fructooligosaccharide achieved the highest score for overall quality acceptance at sensory analyses.

Belorio et al. [3] investigated the utilization of hydrocolloids as a possible approach to optimize the hydration level of gluten-free bread. The authors, evaluating the effect of different hydrocolloids on gluten-free bread's textural aspects, concluded that a thorough investigation of the use of hydrocolloids and starch source mixtures would be required to optimize the gluten-free bread texture.

Arribas et al. [4] studied the effect of cooking on the bioactive compounds content, texture, and color properties of rice/bean-based pasta fortified with carob. The authors' findings revealed that, also after cooking, the fortified pasta maintained its healthy characteristics thanks to the high amount of the detected bioactive compounds. Moreover, this gluten-free pasta did not show the antinutritional factors of bean flour and presented appreciable textural parameters.

Within the framework of gluten-free products, Yu et al. [5] highlighted the importance of gluten detection methods. Currently, several enzyme-linked immunosorbent assay

1

(ELISA) tests are used to detect the trace of gluten, and in their research, the authors observed that three different ELISA test kits often returned values below the detection limits. The authors underlined the importance of developing an accurate analysis method for the detection of gluten traces.

Regarding the fortification process, Tolve et al. [6] enriched pasta with two different levels of grape pomace, an agro-industrial by-product rich in fiber and phenols. The researchers observed an improvement in pasta's nutritional properties and a reduction of the predicted glycemic index. Grape pomace inclusion instigated changes in the cooking and textural properties of pasta. The final product had good overall acceptability from a sensorial point of view. Furthermore, the bread's fortification with two different levels of grape pomace (Tolve et al., 2021) [7] provoked modifications in the rheological properties of the doughs and textural characteristics of the bread samples. The grape pomace inclusion gave rise to the more tenacious and less extensible dough and bread with lower volume. Nevertheless, bread fortification improved the nutritional properties, increasing the total phenol content and the antioxidant capability. The bread samples showed good overall acceptability.

The findings reported in the last two articles suggest that grape pomace represents an interesting ingredient for pasta and baked food fortification, due to the high content in phenols and dietary fiber.

Sissons et al. [8] investigated the amylose content, which is positively correlated with resistant starch, to lower the glycemic index of pasta produced by durum wheat (cv Svevo) by silencing the key genes involved in starch biosynthesis. The results showed that pasta obtained from durum wheat mutants had overall quality acceptability and the starch-branching enzyme IIa's mutation provided a better glycemic response.

Pasini et al. [9] evaluated the technological properties of pasta production with semolina from cv Biensur, produced in zones with different fertility and treated with various rates of N, in comparison with commercial semolina (cv Aureo). The results obtained in this research demonstrated that the technological properties of Biensur semolina correlated to the low fertility zones treated with a high quantity of N. The derived pasta had characteristics similar to the ones obtained by semolina in cv Aureo. The higher amounts of gluten proteins, and the higher glutenin/gliadin ratio in semolina, represent good indexes of technological quality.

Funding: This research received no external funding.

Conflicts of Interest: The author declares no conflict of interest.

References

1. Cappa, C.; Laureati, M.; Casiraghi, M.C.; Erba, D.; Vezzani, M.; Lucisano, M.; Alamprese, C. Effects of Red Rice or Buckwheat Addition on Nutritional, Technological, and Sensory Quality of Potato-Based Pasta. *Foods* **2021**, *10*, 91. [CrossRef] [PubMed]
2. Krupa-Kozak, U.; Drabińska, N.; Rosell, C.M.; Piłat, B.; Starowicz, M.; Jeliński, T.; Szmatowicz, B. High-Quality Gluten-Free Sponge Cakes without Sucrose: Inulin-Type Fructans as Sugar Alternatives. *Foods* **2020**, *9*, 1735. [CrossRef] [PubMed]
3. Belorio, M.; Gómez, M. Effect of Hydration on Gluten-Free Breads Made with Hydroxypropyl Methylcellulose in Comparison with Psyllium and Xanthan Gum. *Foods* **2020**, *9*, 1584. [CrossRef] [PubMed]
4. Arribas, C.; Cabellos, B.; Cuadrado, C.; Guillamón, E.; Pedrosa, M.M. Cooking Effect on the Bioactive Compounds, Texture, and Color Properties of Cold-Extruded Rice/Bean-Based Pasta Supplemented with Whole Carob Fruit. *Foods* **2020**, *9*, 415. [CrossRef]
5. Yu, J.M.; Lee, J.H.; Park, J.; Choi, Y.; Sung, J.; Jang, H.W. Analyzing Gluten Content in Various Food Products Using Different Types of ELISA Test Kits. *Foods* **2021**, *10*, 108. [CrossRef] [PubMed]
6. Tolve, R.; Pasini, G.; Vignale, F.; Favati, F.; Simonato, B. Effect of Grape Pomace Addition on the Technological, Sensory, and Nutritional Properties of Durum Wheat Pasta. *Foods* **2020**, *9*, 354. [CrossRef] [PubMed]
7. Tolve, R.; Simonato, B.; Rainero, G.; Bianchi, F.; Rizzi, C.; Cervini, M.; Giuberti, G. Wheat Bread Fortification by Grape Pomace Powder: Nutritional, Technological, Antioxidant, and Sensory Properties. *Foods* **2021**, *10*, 75. [CrossRef] [PubMed]
8. Sissons, M.; Sestili, F.; Botticella, E.; Mascia, S.; Lafiandra, D. Can Manipulation of Durum Wheat Amylose Content Reduce the Glycaemic Index of Spaghetti? *Foods* **2020**, *9*, 693. [CrossRef] [PubMed]
9. Pasini, G.; Visioli, G.; Morari, F. Is Site-Specific Pasta a Prospective Asset for a Short Supply Chain? *Foods* **2020**, *9*, 477. [CrossRef] [PubMed]

 foods

 MDPI

Article

Analyzing Gluten Content in Various Food Products Using Different Types of ELISA Test Kits

Ja Myung Yu, Jae Hoon Lee, Jong-Dae Park, Yun-Sang Choi, Jung-Min Sung and Hae Won Jang *

Korea Food Research Institute, 245 Nongsaengmyeong-ro, Iseo-myeon, Wanju-gun, Jeollabuk-do 55365, Korea; j.m.yu@kfri.re.kr (J.M.Y.); Leejaehoon@kfri.re.kr (J.H.L.); jdpark@kfri.re.kr (J.-D.P.); kcys0517@kfri.re.kr (Y.-S.C.); jmsung@kfri.re.kr (J.-M.S.)
* Correspondence: hwjkfri@kfri.re.kr; Tel.: +82-63-219-9377; Fax: +82-63-219-9876

Abstract: Gluten is an insoluble protein produced when glutelins and prolamins, which are found in grains such as wheat, barley, and oats, combine to form an elastic thin film. This dietary gluten can cause severe contraction of the intestinal mucous membrane in some people, preventing nutrient absorption. This condition, called celiac disease (CD), affects approximately 1% of the world's population. The only current treatment for patients with CD and similar diseases is lifelong avoidance of gluten. To analyze the gluten content in food, various enzyme-linked immunosorbent assay (ELISA) tests are currently used. In this study, the gluten content in various food products was analyzed using different kinds of ELISA test kits. For gluten-free food, three different ELISA test kits mostly yielded values below the limit of detection. However, gluten was detected at 24.0–40.2 g/kg in bread, 6.5–72.6 g/kg in noodles, and 23.0–86.9 g/kg in different powder food samples. A significant difference ($p < 0.05$) in gluten content was observed for these gluten-containing food products. Reproducibility issues suggest that it is necessary to use several ELISA kits for the accurate detection and quantification of gluten in various food products rather than using one ELISA kit.

Keywords: gluten-free; gluten analysis; ELISA; sandwich method; R5 antibody; G12 antibody; celiac disease

 check for
updates

Citation: Yu, J.M.; Lee, J.H.; Park, J.-D.; Choi, Y.-S.; Sung, J.-M.; Jang, H.W. Analyzing Gluten Content in Various Food Products Using Different Types of ELISA Test Kits. *Foods* 2021, 10, 108. https://doi.org/10.3390/foods10010108

Received: 18 November 2020
Accepted: 26 December 2020
Published: 6 January 2021

1. Introduction

Gluten is a storage protein found in barley, rye, wheat, and their hybrids [1]. The solubility of gluten differs depending on the degree of aggregation. The monomeric protein dissolves in alcohol, and the polymeric forms dissolve in alcoholic solutions under conditions of reduced aggregation [2]. According to its chemical composition or solubility, gluten can be divided into acid-/alkali-soluble glutelin and alcohol-soluble prolamin groups [3,4]. Glutenin in the glutelin group is found in wheat [5]. Gliadin, hordein, secalin, and avenin of the prolamin group are respectively found in wheat, barley, rye, and oats. The prolamin content in gluten is generally determined to be approximately 50% [6,7].

Celiac disease (CD) can be classified as a chronic immune-mediated inflammatory pathology of the small intestine caused by dietary gluten [8]. When CD patients consume gluten-containing food, an autoimmune reaction occurs. The disease causes terrible contraction of the intestinal mucosa, preventing nutrient absorption, and its most common symptoms are diarrhea, anemia, fatigue, and growth retardation [9,10]. Thus, it is essential to maintain a gluten-free diet among these patients, as even trace levels of gluten can damage the mucosal membrane of the small intestine [11]. CD affects approximately 1% of the world's population. The extensive use of yeast and refined grains is one of the most recent causes of such trends [8]. Currently, the only treatment for CD and most other related disorders is lifelong avoidance of gluten in the diet [12,13]. With the increase in the prevalence and the awareness of CD, the gluten-free food industry recorded a 136% growth rate from 2013 to 2015 [14,15].

3

To ensure the safety of gluten-free food for CD patients, they must adhere to a gluten content threshold [16]. According to the Codex Alimentarius Commission (CODEX STAN 118-1979), the Commission Implementing Regulation (EU 828/2014), and the U.S. Food and Drug Administration (FR Doc. 2013-18813), products labeled "gluten-free" must comply with gluten levels of less than 20 mg/kg [17–19]. Products considered "food specifically processed to reduce gluten content" and "low gluten-level" must comply with gluten levels between 20 and 100 mg/kg [20]. Because these values are defined, it is important to provide analytical tools that can accurately and precisely quantify gluten content in various food products [21].

Currently, the Codex Alimentarius Commission proposes enzyme-linked immunosorbent assay (ELISA), as an analytical method, to achieve gluten-free product labeling [22]. This technique includes sandwich methods and competitive polyclonal methods. Most commercial ELISA test kits are based on monoclonal antibodies (R5, Skerritt, G12, and α20) [23–26]. However, ELISA is very expensive, and the reproducibility of the results varies depending on the ELISA test kit type. Differences have been noted in the affinity for gliadin and glutenin of the R5 and the 401/21 antibodies across test kits [27]. Differences have also been observed between values obtained using R5 sandwich and competitive methods for gluten-containing cereals [22]. The effect of food matrices on the detection of gluten is reflected in the differences in recovery [28]. Therefore, whether ELISA is a precise and accurate method to measure gluten content in food remains unclear [29,30]. Furthermore, although many studies have been conducted to confirm the reproducibility of ELISA kits, most of these were performed using wheat flour, and confirmatory studies using a variety of food products on the market are insufficient.

In this study, ELISA methods using R5 and G12 antibodies—against secalin from rye and gliadin from wheat, respectively—[25,31] proposed by the Codex Alimentarius Commission as temporary authorized analytical methods were compared, and calibration was performed for the quantitative analysis of gluten in various food products on the market. The results of the qualitative analysis were compared with those of the quantitative analysis.

2. Materials and Methods

2.1. Materials

Alcoholic beverages, bread, noodles, powdered food items, and snacks, listed in Table 1, were purchased from a local market (Wanju, Korea). Each sample was homogenized in a blender. Samples were stored at −20 °C until analysis.

Table 1. Food products used in analysis.

Type of Food	Product	Label
Bread	Black rice bread	Gluten-free
	Plain bread	Contains gluten
	White rice bread	Gluten-free
Noodles	Buckwheat soba	Contains gluten
	Cellophane noodle	Gluten-free
	Instant noodle	Contains gluten
	Plain noodle	Contains gluten
	Rice noodle	Gluten-free
	Spaghetti noodle	Contains gluten
	Udon noodle	Contains gluten
Powdered food items	Corn starch	Gluten-free
	Green bean powder	Gluten-free
	Potato starch	Gluten-free
	Rice flour	Gluten-free
	Soft wheat flour	Contains gluten
	Strong wheat flour	Contains gluten
	Sugar cane powder	Gluten-free

Table 1. *Cont.*

	Brown rice cereal	Gluten-free
	Brown rice snack	Gluten-free
Snacks	Corn cereal	Gluten-free
	Rice snack	Gluten-free

Gluten in various food products was analyzed using three different sandwich ELISA test kits, as shown in Table 2. The RIDASCREEN Gliadin test kit (R-Biopharm AG, Darmstadt, Germany), the Veratox for Gliadin R5 test kit (Neogen, Lansing, MI, USA.), and the AgraQuant Gluten G12 test kit (Romer Labs, Runcorn, U.K.) were used for quantitative analysis. Using the qualitative test kit AgraStrip Gluten G12 test kit (Romer Labs, Runcorn, UK), the gluten content was analyzed and compared with the results of quantitative analysis. The analytical method for gluten detection and analysis followed protocols provided by each test kit manufacturer. All food samples were analyzed in triplicate.

Table 2. Characteristics of commercial ELISA test kits.

Test Kit	Manufacturer	Format	Antibody	Target Protein
RIDASCREEN Gliadin	R-Biopharm	Sandwich	R5	Gliadin
Veratox for Gliadin R5	Neogen	Sandwich	R5	Gliadin
AgraQuant Gluten G12	Romer Labs	Sandwich	G12	Gluten
AgraStrip Gluten G12	Romer Labs	Lateral flow device (LFD)	G12	Gluten

2.2. Quantitative Analysis of Gluten Using RIDASCREEN Gliadin Test

The analytical protocol followed the ELISA test kit manufacturer's instructions precisely. The sample (0.5 g) was placed into a 50 mL centrifuge tube and incubated for 40 min at 50 °C by adding 2.5 mL of the cocktail solution. Subsequently, 80% ethanol (7.5 mL) was added and mixed for 60 min to extract gluten. Samples were centrifuged for 10 min at $2500 \times g$ at room temperature. Three independent extraction procedures for each food sample were performed with triplicate measurements.

Next, 100 µL of each blank, standard, and sample solution was added into the wells and incubated for 30 min at room temperature, after which the standard and the sample solutions were removed from the wells. All wells were washed with a wash buffer three times. Thereafter, 100 µL of the conjugate was added to the wells and incubated for 30 min. Next, the conjugate was removed, and the wells were washed three times. Thereafter, 50 µL of the substrate and 50 µL of chromogen were added to each well and incubated for 30 min in the dark. Finally, 100 µL of the stop solution was added to measure the absorbance at 450 nm.

2.3. Quantitative Analysis of Gluten Using Veratox for Gliadin R5 kit

For samples that were not subjected to the heat treatment process, 1.0 g of the sample and the extraction additive were placed in a centrifuge tube (Fisher Scientific, Pittsburgh, PA, USA). Subsequently, 10 mL of 60% ethanol was added and mixed for 10 min. After centrifugation for 10 min at $2500 \times g$, 100 µL of the upper layer of the extract was put into the tube, and 4.9 mL of the sample extract dilution solution (phosphate buffered saline, PBS, Sigma, MO, USA) was added to dilute each sample at a 1:50 ratio. The diluted samples were analyzed within 2–3 h.

Samples (0.25 g) for the heat treatment process were put into a tube, and 2.5 mL of the cocktail solution was added (if the samples contained buckwheat, chestnut, and tannin, an extraction additive was added to prevent disruption of analysis due to polyphenols) [22]. The mixture was homogenized for 30 s and incubated at 50 °C for 10 min. Three minutes later, 7.5 mL of 80% ethanol was added, and the mixture was shaken for 1 h. After extraction, the sample was centrifuged for 10 min at 2500× *g*. PBS (2.3 mL) was added to 200 μL of the upper layer to dilute the sample at a 1:12.5 ratio.

Approximately 150 μL of a blank, standard, and samples was injected into the red-marked mixing well, and 100 μL of each aliquot was moved into the antibody-coated well, and then incubation was performed for 10 min, after which the red-marked mixing well was removed from the plate. The standard and the sample solution in the antibody-coated well were removed, and the wells were washed five times with a wash buffer. After adding 100 μL of the conjugate to the well, the plate was incubated for 10 min, and then the conjugate was removed and the well washed five times. Next, 100 μL of the substrate was added to the well and incubated for 10 min, and then 100 μL of the stop solution was added to measure the absorbance at 650 nm (VersaMax™ microplate ELISA reader, Molecular Device, CA, USA).

2.4. Quantitative Analysis of Gluten Using AgraQuant Gluten G12 kit

A 0.25 g specimen was placed in a tube, and 2.5 mL of an extension solution was added. The mixture was incubated at 50 °C for 40 min, and 80% ethanol (7.5 mL, Merch, Darmstadt, Germany) was added with a rotator for 60 min for extraction. The extracts were centrifuged for 10 min at 2000× *g*.

The wells were then washed with the wash buffer five times. Thereafter, 100 μL of the conjugate was added to each well and incubated for 20 min and then removed. The washing step was repeated five times. Next, 100 μL of the substrate was added to the well and incubated for 20 min in the dark. Finally, 100 μL of the stop solution was added to each well to measure the absorbance at 450 nm.

2.5. Qualitative Analysis of Gluten Using AgraStrip Gluten G12 kit

A 0.2 g sample was put into the extraction tube, and 2.5 mL of the extension buffer was added. About 100 μL of the extract was put into the dilution tube, and the dilution buffer was added up to the mark of 5 mg/kg. After dipping the test strip vertically, we waited for 45 s for the solution to rise to the flow level line. The test strip was removed from the tube, and the result was checked 10 min later. If a single blue line appeared in the result zone, it was considered a negative result, but if blue and red lines appeared, it was considered a positive result. If a positive reaction occurred at 5 mg/kg, the dilution buffer was added up to the 10 mg/kg and 20 mg/kg marks, and the same process was repeated.

2.6. Statistical Analysis

Statistical analysis was performed using IBM SPSS Statistics 23 (IBM, Armonk, NY, USA). Data were analyzed by one-way ANOVA and Duncan's multiple range test for investigating significant differences ($p < 0.05$).

3. Results

3.1. Calibration of ELISA Test Kits

Gluten usually includes gliadin and glutenin at a ratio of 1:1 [32,33]. In the case of the RIDASCREEN gliadin kit and the Veratox for gliadin R5 kit, twice the quantitation value of gliadin was calculated as the approximate content of gluten. The AgraQuant Gluten G12 kit indicates the detected gluten content. For quantitative analysis, standard calibration curves of five points were obtained using each ELISA kit. The limit of detection (LOD) and the limit of quantitation (LOQ) were validation data specified by the manufacturers. The results and data are shown in Table 3.

Table 3. Linearity and sensitivity of sandwich ELISA test kits.

Test Kit	Linearity			Sensitivity	
	Linear Range (ng/mL)	Calibration Curve	R^2	LOD (mg/kg)	LOQ (mg/kg)
RIDASCREEN gliadin kit	5–40 5–80	Linear Quadratic	0.9175 0.9953	1.0	5.0
Veratox for gliadin R5	5–40 5–80	Linear Quadratic	0.9966 0.9988	5.0	5.0
AgraQuant Gluten G12	5–100 5–200	Linear Quadratic	0.9874 0.9958	2.0	4.0

LOD and LOQ mean limit of detection and limit of quantification, respectively.

Calibration of RIDASCREEN gliadin kit based on the R5 antibody was performed as follows. Linearity was confirmed in the concentration range of 5-40 µg/mL. The correlation coefficient for the linearity of the four points was $R^2 > 0.91$. For the concentration range of 5-80 µg/mL, the correlation coefficient for the quadratic of five points was $R^2 > 0.99$. More than 1.0 mg/kg of gluten could be detected, and the LOQ was 5.0 mg/kg.

The linearity of Veratox for gliadin R5 kit based on the R5 antibody was found in the 5–40 µg/mL concentration range. The correlation coefficient of the four points for linearity was $R^2 > 0.99$. In the case of the concentration range of 5-80 µg/mL, the correlation coefficient for the quadratic of five points was $R^2 > 0.99$. The LOD and the LOQ were estimated at 5.0 mg/kg, respectively.

The linearity of AgraQuant gluten G12 kit based on the G12 antibody was found in the 5–100 µg/mL concentration range. The correlation coefficient for the four points was $R^2 > 0.98$. For the concentration range of 5-200 µg/mL, the correlation coefficient for the quadratic of five points was $R^2 > 0.99$. The estimated LOD and LOQ were 2.0 mg/kg and 4.0 mg/kg, respectively.

From the linearity results of the absorbance readings, a quadratic regression was used in all samples in this study.

3.2. Results of Qualitative Analysis of Gluten in Products

The qualitative analysis results using the AgraStrip Gluten G12 Kit are shown in Table 4. Gluten detection was performed for 21 types of food products. Among breads, only plain bread made of wheat flour gave a positive result. Among noodles, buckwheat soba, plain noodles, instant noodles, spaghetti noodles, and udon noodles tested positive. Among the powders, strong and soft wheat flour tested positive. All snacks tested negative. In total, only eight samples were found to contain gluten.

Table 4. Results of qualitative analysis of gluten using AgraStrip Gluten G12 test kit.

Type of Food	Product	Test Result	
		5 mg/kg	10 mg/kg
Bread	Black rice bread Plain bread White rice bread	− + −	− + −
Noodles	Buckwheat soba Cellophane noodle Instant noodle Plain noodle Rice noodle Spaghetti noodle Udon noodle	+ − + + − + +	+ − + + − + +

Table 4. *Cont.*

Powder	Corn starch	-	-
	Green bean powder	-	-
	Potato starch	-	-
	Rice flour	-	-
	Soft wheat flour	+	+
	Strong wheat flour	+	+
	Sugar cane powder	-	-
Snacks	Brown rice cereal	-	-
	Brown rice snack	-	-
	Corn cereal	-	-
	Rice snack	-	-

If the result is positive, it is expressed as (+); otherwise, it is denoted as (-).

3.3. Results of Quantitative Analysis of Gluten in Gluten-Containing Products

The results of gluten content in eight samples containing gluten obtained using three sandwich ELISAs are shown in Figure 1. Gluten was detected in all the samples. Supplementary Table S1 shows the average and the relative standard deviations. In the case of powders, the highest content was 51.2–86.9 g/kg for strong wheat flour. Soft wheat flour was found to contain 23.0–47.3 g/kg of gluten. Flour is classified according to its gluten content. If the gluten content is high, it is classified as a strong flour; otherwise, it is classified as soft. Of the noodles, buckwheat soba showed the highest gluten content at 43.2–72.6 g/kg, followed by plain noodles (43.7–53.0 g/kg), instant noodles (12.0–35.3 g/kg), udon noodles (6.5–30.3 g/kg), and spaghetti noodles (3.7–20.9 g/kg). Plain bread gave a value of 24.0–40.2 g/kg. A significant difference ($p < 0.05$) was noted when quantification was undertaken using three different ELISA test kits for gluten-rich food.

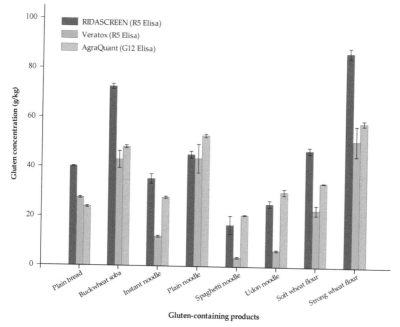

Figure 1. Gluten concentration in gluten-containing products measured using three types of sandwich ELISA.

3.4. Results of Gluten Content in Gluten-Free Products

The results of gluten content in 13 types of gluten-free samples are shown in Table 5. Bread, noodles, and snacks had values below the LOD or the LOQ. Among powder products, when analyzed using AgraQuant kit, a small amount (5.6 mg/kg) of gluten was detected only in green bean powder.

Table 5. Results for gluten-free products using three types of sandwich ELISA test kits.

Type of food	Product	Gluten Concentration (mg/kg)		
		RIDASCREEN (R5 ELISA)	Veratox (G5 ELISA)	AgraQuant (G12 ELISA)
Bread	Black rice bread	Below LOD	Below LOD	Below LOD
	White rice bread	Below LOD	Below LOD	Below LOD
Noodles	Cellophane noodle	Below LOD	Below LOD	Below LOD
	Rice noodle	Below LOD	Below LOD	Below LOD
Powder	Corn starch	Below LOD	Below LOD	Below LOD
	Green bean powder	Below LOQ	Below LOD	5.6 ± 0.4
	Potato starch	Below LOD	Below LOD	Below LOD
	Rice flour	Below LOD	Below LOD	Below LOD
	Sugar cane powder	Below LOD	Below LOD	Below LOD
Snacks	Brown rice cereal	Below LOD	Below LOD	Below LOD
	Brown rice snack	Below LOD	Below LOD	Below LOD
	Corn cereal	Below LOD	Below LOD	Below LOD
	Rice snack	Below LOQ	Below LOD	Below LOD

All values are denoted as mean ± standard deviation (*n* = 3).

4. Discussion

In this study, the reproducibility of several commercial ELISA test kits for the quantification of gluten content was assessed. The RIDASCREEN and the Veratox test kits employ the affinity of the R5 antibody for gliadin, whereas AgraQuant employs that of the G12 antibody for gliadin. All three types of test kits were used in the sandwich method. For the RIDASCREEN gliadin test kit and the Veratox for Gliadin R5 test kit, the correlation coefficients for quadratic regression in the concentration range of 5–80 ng/mL, were $R^2 > 0.99$ and $R^2 > 0.99$, respectively. The LODs were 1.0 mg/kg and 5.0 mg/kg, respectively, and the LOQs were 5.0 mg/kg for both. For the AgraQuant Gluten G12 test kit, the correlation coefficient for quadratic in the concentration range of 5 to 200 ng/mL was $R^2 > 0.99$. The LODs and the LOQs were 2.0 and 4.0 mg/kg, respectively. For the qualitative analysis of gluten using the AgraStrip Gluten G12 kit, the gluten-free food tested negative for gluten. Using three different ELISA test kits for quantitative analysis, the obtained results were mostly below the LOD. Some gluten-free powders, such as green bean powder, had gluten contents of 5.6 mg/kg. This may due to the contamination with flour during production. The plant where the powder production process is carried out also produces wheat flour, corn starch, green bean powder, and sugarcane powder, leading to a potential cross-contamination of the products. However, according to CODEX, European Commission Regulation, and the U.S. Food and Drug Administration, products with a gluten content of less than 20 mg/kg can be labeled "gluten-free" [18,19].

In contrast, the gluten contents of gluten-rich products were 23.0–86.9 g/kg for flour, 3.7–72.6 g/kg for noodles, and 24.0–40.2 g/kg for bread. For these gluten-rich products, when quantification was carried out using the three different ELISA test kits, a significant difference (*p* < 0.05) in gluten content was observed. This is due to differences in the antibody characteristics and the extract solutions of the ELISA test kits. Various food matrices remain difficult to analyze owing to either interference of antibody binding by the food matrix or cross-reactivity [22]. The results depend on the extraction method when using a cocktail solution containing 2-mercaptoethanol [34]. Several previous studies have reported similar results to ours. Scherf [35] reported that seven commercial ELISA test kits showed different gluten assay results in wheat products, especially gluten-free wheat. In addition, other studies using ELISA kits reported gluten contents higher than the stipulated threshold for gluten-free products. Likewise, Bruins Slot et al. [22] also reported similar results revealing that oat flour (a gluten-free labeled product) had a gluten content of more than 20 mg/kg and that the measurement error is large between the different commercial gluten kits used to measure the gluten content in foods. They concluded that it was due to the difference in the ability to extract gluten from the food matrix, the difference in the sensitivity of the used antibody, and in the reference material of the kits. As a result, the analysis of gluten content using such kits has some drawbacks, and thus an accurate analysis through a new analysis technique is required.

5. Conclusions

In summary, using three commercial ELISA test kits for measuring gluten content in various food products, calibration and quantitative analyses were conducted. The results showed that the reproducibility of the three kits was low. These kits had to accurately detect gluten in mg/kg units because the standard level for gluten-free products is 20 mg/kg. However, the experimental results confirmed an error in the g/kg units. Therefore, methods of extraction or device analysis that utilize precision analysis devices should be studied to ensure safe products for CD patients.

Supplementary Materials: The following are available online at https://www.mdpi.com/2304-8158/10/1/108/s1. Table S1, Concentration of gluten in gluten-containing products using three types of sandwich ELISA test kits.

Author Contributions: Conceptualization, J.M.Y. and H.W.J.; methodology, J.M.Y. and H.W.J.; software, J.-D.P. and Y.-S.C.; validation, J.H.L., J.-M.S., and H.W.J.; formal analysis, J.M.Y. and Y.-S.C.; investigation, J.-D.P. and J.M.Y.; data curation, J.M.Y., J.H.L., and H.W.J.; writing—original draft preparation, J.M.Y., J.H.L., and H.W.J.; writing—review and editing, Y.-S.C. and H.W.J.; visualization, J.H.L.; supervision, H.W.J.; All authors have read and agreed to the published version of the manuscript.

Funding: This research was funded by Korea Food Research Institute (Project No. E0193115-02) and Agri-FoodExport Business Model development Program (319090-3) by the Ministry of Agriculture, Food and Rural Affairs.

Institutional Review Board Statement: Not applicable.

Informed Consent Statement: Not applicable.

Data Availability Statement: The data presented in this study are available in the article and the supplementary material.

Conflicts of Interest: The authors declare no conflict of interest.

References

1. Biesiekierski, J.R. What is gluten? *J. Gastroenterol. Hepatol.* **2017**, *32*, 78–81. [CrossRef]
2. Mena, M.C.; Lombardía, M.; Hernando, A.; Méndez, E.; Albar, J.P. Comprehensive analysis of gluten in processed foods using a new extraction method and a competitive ELISA based on the R5 antibody. *Talanta* **2012**, *91*, 33–40. [CrossRef]
3. Diaz-Amigo, C.; Popping, B. Accuracy of ELISA detection methods for gluten and reference materials: A realistic assessment. *J. Agric. Food Chem.* **2013**, *61*, 5681–5688. [CrossRef]
4. Sharma, G.M.; Khuda, S.E.; Pereira, M.; Slate, A.; Jackson, L.S.; Pardo, C.; Williams, K.M.; Whitaker, T.B. Development of an incurred cornbread model for gluten detection by immunoassays. *J. Agric. Food Chem.* **2013**, *61*, 12146–12154. [CrossRef]
5. Wieser, H. Chemistry of gluten proteins. *Food Microbiol.* **2007**, *24*, 115–119. [CrossRef] [PubMed]
6. Koehler, P.; Schwalb, T.; Immer, U.; Lacorn, M.; Wehling, P.; Don, C. AACCI approved methods technical committee report: Collaborative study on the immunochemical determination of intact gluten using an R5 sandwich ELISA. *Cereal Food World* **2013**, *58*, 36–40. [CrossRef]

7. Scherf, K.A.; Poms, R.E. Recent developments in analytical methods for tracing gluten. *J. Cereal Sci.* **2016**, *67*, 112–122. [CrossRef]
8. Gobbetti, M.; Pontonio, E.; Filannino, P.; Rizzello, C.G.; De Angelis, M.; Di Cagno, R. How to improve the gluten-free diet: The state of the art from a food science perspective. *Food Res. Int.* **2018**, *110*, 22–32. [CrossRef] [PubMed]
9. Ribeiro, M.; Nunes, F.M.; Rodriguez-Quijano, M.; Carrillo, J.M.; Branland, G.; Igrejas, G. Next-generation therapies for celiac disease: The gluten-targeted approaches. *Trends Food Sci. Technol.* **2018**, *75*, 56–71. [CrossRef]
10. Goel, G.; Tye-Din, J.A.; Qiao, S.-W.; Russell, A.K.; Mayassi, T.; Ciszewski, C.; Sarna, V.K.; Wang, S.; Goldstein, K.E.; Dzuris, J.L. Cytokine release and gastrointestinal symptoms after gluten challenge in celiac disease. *Sci. Adv.* **2019**, *5*, eaaw7756. [CrossRef]
11. Sapone, A.; Bai, J.C.; Ciacci, C.; Dolinsek, J.; Green, P.H.; Hadjivassiliou, M.; Kaukinen, K.; Rostami, K.; Sanders, D.S.; Schumann, M.; et al. Spectrum of gluten-related disorders: Consensus on new nomenclature and classification. *BMC Med.* **2012**, *10*, 1–12. [CrossRef] [PubMed]
12. Lynch, K.M.; Coffey, A.; Arendt, E.K. Exopolysaccharide producing lactic acid bacteria: Their techno-functional role and potential application in gluten-free bread products. *Food Res. Int.* **2018**, *110*, 52–61. [CrossRef] [PubMed]
13. Bascunan, K.A.; Vespa, M.C.; Araya, M. Celiac disease: Understanding the gluten-free diet. *Eur. J. Nutr.* **2017**, *56*, 449–459. [CrossRef] [PubMed]
14. Reilly, N.R. The gluten-free diet: Recognizing fact, fiction, and fad. *J. Pediatr.* **2016**, *175*, 206–210. [CrossRef] [PubMed]
15. Morreale, F.; Angelino, D.; Pellegrini, N. Designing a score-based method for the evaluation of the nutritional quality of the gluten-free bakery products and their gluten-containing counterparts. *Plant Foods Hum. Nutr.* **2018**, *73*, 154–159. [CrossRef]
16. Lacorn, M.; Scherf, K.; Uhlig, S.; Weiss, T. Determination of gluten in processed and nonprocessed corn products by qualitative R5 immunochromatographic dipstick: Collaborative study, first action 2015.16. *J. AOAC Int.* **2016**, *99*, 730–737. [CrossRef]
17. Hochegger, R.; Mayer, W.; Prochaska, M. Comparison of R5 and G12 antibody-based ELISA used for the determination of the gluten content in official food samples. *Foods* **2015**, *4*, 654–664. [CrossRef]
18. CODEX, S. STAN 118-1979. *Standard for Foods for Special Dietary Use for Persons Intolerant to Gluten*; Food and Agriculture Organization of the United Nations: Quebec, QC, Canada; World Health Organization: Geneva, Switzerland, 2008.
19. Commission, E. Commission implementing regulation (EU) No 828/2014, requirements for the provision of information to consumers on the absence or reduced presence of gluten in food. *Off. J. Eur. Union L.* **2014**, *228*, 5–8.
20. Verma, A.K.; Gatti, S.; Galeazzi, T.; Monachesi, C.; Padella, L.; Baldo, G.D.; Annibali, R.; Lionetti, E.; Catassi, C. Gluten contamination in naturally or labeled gluten-free products marketed in Italy. *Nutrients* **2017**, *9*, 115. [CrossRef]
21. Bugyi, Z.; Török, K.; Hajas, L.; Adonyi, Z.; Popping, B.; Tömösközi, S. Comparative study of commercially available gluten ELISA kits using an incurred reference material. *Qual. Assur. Saf. Crop* **2013**, *5*, 79–87. [CrossRef]
22. Bruins Slot, I.D.; Bremer, M.G.; van der Fels-Klerx, I.; Hamer, R.J. Evaluating the performance of gluten ELISA test kits: The numbers do not tell the tale. *Cereal Chem.* **2015**, *92*, 513–521. [CrossRef]
23. Sharma, G.M. Immunoreactivity and detection of wheat proteins by commercial ELISA kits. *J. AOAC Int.* **2012**, *95*, 364–371. [CrossRef] [PubMed]
24. Skerritt, J.H.; Hill, A.S. Monoclonal antibody sandwich enzyme immunoassays for determination of gluten in foods. *J. Agric. Food Chem.* **1990**, *38*, 1771–1778. [CrossRef]
25. Moron, B.; Bethune, M.T.; Comino, I.; Manyani, H.; Ferragud, M.; Lopez, M.C.; Cebolla, Á.; Khosla, C.; Sousa, C. Toward the assessment of food toxicity for celiac patients: Characterization of monoclonal antibodies to a main immunogenic gluten peptide. *PLoS ONE* **2008**, *3*, e2294. [CrossRef] [PubMed]
26. Spaenij-Dekking, L.; Kooy-Winkelaar, E.; Nieuwenhuizen, W.; Drijfhout, J.; Koning, F. A novel and sensitive method for the detection of T cell stimulatory epitopes of α/β-and γ-gliadin. *Gut* **2004**, *53*, 1267–1273. [CrossRef]
27. Allred, L.K.; Ritter, B.W. Recognition of gliadin and glutenin fractions in four commercial gluten assays. *J. AOAC Int.* **2010**, *93*, 190–196. [CrossRef]
28. Scharf, A.; Kasel, U.; Wichmann, G.; Besler, M. Performance of ELISA and PCR methods for the determination of allergens in food: An evaluation of six years of proficiency testing for soy (Glycine max L.) and wheat gluten (Triticum aestivum L.). *J. Agric. Food Chem.* **2013**, *61*, 10261–10272. [CrossRef]
29. Lupo, A.; Roebuck, C.; Walsh, A.; Mozola, M.; Abouzied, M. Validation study of the Veratox R5 Rapid ELISA for detection of gliadin. *J. AOAC Int.* **2013**, *96*, 121–132. [CrossRef]
30. Koerner, T.B.; Abbott, M.; Godefroy, S.B.; Popping, B.; Yeung, J.M.; Diaz-Amigo, C.; Roberts, J.; Taylor, S.L.; Baumert, J.L.; Ulberth, F. Validation procedures for quantitative gluten ELISA methods: AOAC allergen community guidance and best practices. *J. AOAC Int.* **2013**, *96*, 1033–1040. [CrossRef]
31. Kanerva, P.M.; Sontag-Strohm, T.S.; Ryöppy, P.H.; Alho-Lehto, P.; Salovaara, H.O. Analysis of barley contamination in oats using R5 and ω-gliadin antibodies. *J. Cereal Sci.* **2006**, *44*, 347–352. [CrossRef]
32. Barak, S.; Mudgil, D.; Khatkar, B. Relationship of gliadin and glutenin proteins with dough rheology, flour pasting and bread making performance of wheat varieties. *LWT Food Sci. Technol.* **2013**, *51*, 211–217. [CrossRef]
33. Don, C.; Halbmayr-Jech, E.; Rogers, A.; Koehler, P. AACCI Approved Methods Technical Committee report: Collaborative study on the immunochemical quantitation of intact gluten in rice flour and rice-based products using G12 sandwich ELISA. *Cereal Food World* **2014**, *59*, 187–193. [CrossRef]

34. Geng, T.; Westphal, C.D.; Yeung, J.M. Detection of gluten by commercial test kits: Effects of food matrices and extraction procedures. *ACS Symp. Ser.* **2008**, *1001*, 462–475.
35. Scherf, K.A. Gluten analysis of wheat starches with seven commercial ELISA test kits—Up to six different values. *Food Ana. Meth.* **2017**, *10*, 234–246. [CrossRef]

foods

MDPI

Article

Effects of Red Rice or Buckwheat Addition on Nutritional, Technological, and Sensory Quality of Potato-Based Pasta

Carola Cappa [1],*, Monica Laureati [1], Maria Cristina Casiraghi [1], Daniela Erba [1], Maurizio Vezzani [2], Mara Lucisano [1] and Cristina Alamprese [1]

1   Dipartimento di Scienze per Gli Alimenti, la Nutrizione e l'Ambiente, Università Degli Studi di Milano, Via G. Celoria, 2-20133 Milano, Italy; monica.laureati@unimi.it (M.L.); maria.casiraghi@unimi.it (M.C.C.); daniela.erba@unimi.it (D.E.); mara.lucisano@unimi.it (M.L.); cristina.alamprese@unimi.it (C.A.)
2   Zini Prodotti Alimentari S.p.A, Via Libertà 36, 20090 Cesano Boscone, Milano, Italy; mv@pastazini.it
*   Correspondence: carola.cappa@unimi.it; Tel.: +39-02-5031-9179; Fax: +39-5031-9190

Abstract: This work investigates the effects of red rice (R) or buckwheat (B) flour addition on nutritional, technological, and sensory quality of potato-based pasta (gnocchi). Three gluten-free (GF) and three conventional (C) samples were produced in an industrial line without any addition or with 20% R or B. R and B addition significantly ($p < 0.05$) reduced starch content and increased fat amount and ready digestible starch fraction (potential higher glycemic impact). R addition significantly ($p < 0.05$) worsened GF pasta structure, increasing solid loss in cooking water (5.4 ± 1.2 vs. 4.1 ± 0.5 g/100 g pasta) and reducing product firmness (408 ± 13 vs. 108 ± 2 N). B addition resulted in intermediate consistency (243 ± 8 N), despite the highest total fiber content and weight increase during cooking. Similar trends were found in C samples, indicating a better texturizing capacity of B in comparison to R. Samples without any addition were the most liked (C = 67.4 and GF = 60.6). Texture was the major contributor to liking: uniform structure and firm texture were positive predictors of liking, whereas a granular and coarse matrix contributed negatively. The outcomes of this research can be useful in developing GF potato-based pasta for consumers focused on healthier foods and for industries willing to better valorize their products.

Keywords: dumpling; gnocchi; gluten free pasta; fiber content; starch digestibility; cooking behavior; color; texture; liking predictors; consumer acceptability

Citation: Cappa, C.; Laureati, M.; Casiraghi, M.C.; Erba, D.; Vezzani, M.; Lucisano, M.; Alamprese, C. Effects of Red Rice or Buckwheat Addition on Nutritional, Technological, and Sensory Quality of Potato-Based Pasta. *Foods* **2021**, *10*, 91. https://doi.org/10.3390/foods10010091

Received: 2 December 2020
Accepted: 26 December 2020
Published: 5 January 2021

Publisher's Note: MDPI stays neutral with regard to jurisdictional claims in published maps and institutional affiliations.

1. Introduction

The increase in celiac disease and other allergic reactions to gluten has opened new market opportunities for pasta producers, especially in the sector of fresh products, in which the gluten-free offering is still limited. The removal of gluten represents a challenge for good quality products, because it is responsible for the well appreciated pasta structure. In gluten-free pasta, structure is assured mainly by starch, whose gelatinization degree plays an important manufacturing role [1,2]. In fact, the use of pregelatinized starch ingredients allows the application of a standard pasta production process, whereas non-pregelatinized starch sources require gelatinization to occur during processing. Usually, proteins, hydrocolloids, and emulsifiers are also included in the formulation to improve gluten-free dough workability and quality of the final product [3]. Nutritional properties of gluten-free pasta are not comparable to those of conventional products, because of the reduced levels of dietary fiber, resistant starch, and protein, with higher glycemic index and starch digestion rate [4].

Among fresh pasta products, potato dumplings are very popular in many countries and they are prepared in different way [5]. The Italian version is called "gnocchi"; it mainly consists of potato (fresh or dehydrated), to which wheat flour and salt are added; eventually, eggs, emulsifiers, and preservatives can be used in the recipe [6]. In order to improve nutritional properties of conventional and gluten-free gnocchi, different strategies

13

have been proposed, such as the addition of quinoa and amaranth flours [7], green coffee extract [8], navy bean flour, and meat [9]. In this context, the enrichment with red rice or buckwheat flours could also have a positive effect.

Red rice is a pigmented variety of rice (*Oryza sativa* L.) with beneficial health effects due to the antioxidant activity of bioactive compounds such as phenolic compounds, anthocyanins, and proanthocyanidins, which are associated with protection against chronic diseases [10]. A number of papers demonstrated the anti-oxidant, anti-diabetic, anti-hyperlipidemic, and anti-cancer activity of pigmented rice varieties, which are thus gaining popularity among consumers. However, texture and palatability of pigmented rice are poor and, thus, the consumers' acceptance is low [11], justifying the limited number of studies on red rice enriched pasta [12,13].

Buckwheat (*Fagopyrum* spp.) is a pseudocereal belonging to the family of the *Polygonaceae*, with a more balanced amino acid composition and, thus, biological value higher than that of most cereals. It can safely be consumed by people suffering from celiac disease and it is rich in constituents important for human health, such as dietary fiber, antioxidants, minerals, and vitamins [14]. Moreover, buckwheat has a high level of resistant starch (27–33.5%), which can help in modulating blood glucose and lipid levels, regulating intestinal microbiota, and reducing obesity [15]. The use of buckwheat in bread, cookies, and pasta formulations, both conventional and gluten-free, has been extensively studied [14,15], however the effects of its addition in gnocchi recipes have not yet been evaluated.

The aim of this study was to investigate the effects of red rice or buckwheat flour addition (20%) on the nutritional, technological, and sensory qualities of conventional and gluten-free quick-frozen gnocchi, produced by a turbo-cooking technology. This thermal technology, patented by VOMM Impianti e Processi S.p.A. (Rozzano, Italy), causes the starch gelatinization during gnocchi production, allowing to obtain a good final product without including pregelatinized ingredients in the formulation [6].

2. Materials and Methods

Three gluten-free (GF) and three conventional (C) samples of potato-based pasta (i.e., gnocchi) were produced without or with 20% addition of wholemeal red rice (R; Distretto rurale "Riso e Rane", Cassinetta di Lugagnano, Italy) or wholemeal buckwheat (B; Molino Filippini S.r.l., Teglio, Italy) flour (Figure 1). All of the other ingredients were provided by Zini Prodotti Alimentari S.p.A. (Cesano Boscone, Italy).

| GF | GFR | GFB |
| C | CR | CB |

Figure 1. Gluten free (GF) and conventional (C) potato-based pasta samples without or with 20% addition of red rice (R) or buckwheat (B) flour.

The potato-based pasta recipe was defined according to literature data [6] and Zini Prodotti Alimentari S.p.A. (Cesano Boscone, Italy) experience. GF reference sample was made of water, rice flour, dehydrated potato (24 g/100 g), corn flour, and salt; C reference sample contained water, semolina flour, dehydrated potato (33 g/100 g), and salt. All samples were produced by Zini Prodotti Alimentari S.p.A. (Cesano Boscone, Italy) in an industrial line by a turbo-cooking technology (VOMM® Impianti e Processi S.p.A., Rozzano, Italy) followed by individual quick-freezing at −35 °C [6]. Samples were stored at −18 °C till their characterization.

ABTS (2,2-azino-bis(3-ethylbenzothiazoline-6-sulphonic acid) diammonium salt, A1888), Trolox (6-hydroxy-2,5,7,8-tetramethylchroman-2-carboxylic acid, 238,813), potassium persulphate (dipotassium peroxydisulphate, P-5592), pepsin (P7000; ≥250 U/mg), pancreatin (P7545; 8xUSP), invertase (I4504; ≥300 U/mg) and amyloglucosidase (A7095; ≥260 U/mL), were purchased from Sigma Chemical Co. (St. Louis, MO, USA) and chemicals at analytical grade from Merck KGaA (Darmstadt, Germany).

2.1. Pasta Cooking Conditions

For nutritional and technological evaluation, pasta samples were cooked in boiling unsalted tap water (1:10 pasta:water ratio) for their optimal cooking time (OCT) and drained for 1 min. For sensory evaluation, samples were cooked at OCT in salted tap water (salt: 10 g/L) in order to make the evaluation more similar to common consumption. The OCT of each sample was defined according to preliminary sensory tests: GF and C, 120 s; CR, 100 s; GFR, GFB, and CB, 90 s.

2.2. Nutritional Quality Evaluation

Raw gnocchi samples were ground in a mixer (Bimby VM 2200, Thermomix, Worwerk, Wuppertal, Germany) for 5 min at maximum speed and characterized in terms of composition. Moisture and ash content were evaluated by the official gravimetric methods, lipids were extracted with a mixture of ethyl ether and petroleum ether (2:1) using a Soxhlet apparatus and nitrogen content was detected according to the Kjeldahl method [16]. Proteins were calculated using 6.25 as the nitrogen/protein conversion factor. HPLC with an ion exchange column combined with a pulsed amperometry detection system was used to evaluate soluble sugars [17]. Soluble and insoluble fibers were evaluated by an enzymatic-gravimetric procedure [16,18]. The antioxidant capacity was measured by the ABTS˙ assay and expressed as Trolox equivalents (Trolox standard concentrations were 2–22 μmol/L; calibration curve, $r = 0.994$) [19]. All of the evaluations were run in triplicate.

Cooked gnocchi were ground by using a screw-type kitchen grinder, for the analysis of in vitro starch digestibility [20,21]. The method measures the rate of digestion through a series of proteolytic and amylolytic enzymatic attacks under controlled conditions of temperature, pH, viscosity, and stirring speed, simulating the different digestive steps that take place in vivo. Based on the HPLC analysis of the glucose released after 20 min (G20) and 120 min (G120), the fractions of ready (RDS) and slowly (SDS) digestible starch were calculated. The sum of RDS and SDS is indicative of the starch digestible in the small intestine and it is defined as available starch (AvSt). The ratios of RDS and SDS over AvST were also calculated. The digestibility tests were conducted in duplicate and repeated 4 times ($n = 8$).

2.3. Technological Quality Evaluation

Color of R and B was evaluated by using a Minolta Chroma Meter II (Minolta, Osaka, Japan) with standard illuminant C, on flours (approximately 30 g) levelled in petri dishes. Results were expressed in the CIE L*a*b* space as L* (lightness; from black (0) to white (100)), a* (from green (−) to red (+)), and b* (from blue (−) to yellow (+)) values. The particle size distribution of the flour samples (50 g) was evaluated by means of the analytical sieve shaker Octagon Digital (Endecotts Ltd., London, UK), by using 4 certified sieves (openings: 90, 125, 250, and 500 μm). Five fractions were collected after sieving for 10 min at amplitude 6 in the presence of 3 plastic spheres (diameter: 3.0 cm) on each sieve, to make the sifting of the fine particles easier. Triplicate measurements were performed for each sample and results were expressed as g/100 g for each fraction.

Raw gnocchi were characterized in terms of weight, surface color (using a colorimeter Minolta Chroma Meter II; Minolta, Osaka, Japan), and geometrical indices (by image analysis according to literature data [2]). Sample images were acquired at 300 dpi resolution using a flatbed scanner (HP SCANJET8300; Hewlett-Packard Development Company, Palo Alto, CA, USA) and covering gnocchi with a black box to amplify the contrast between the objects and the background and to prevent light losses. Images were processed using a dedicated software (Image Pro-Plus v. 4.5.1.29, Media Cybernetics Inc., Rockville, MD, USA) in order to measure sample area, width, and length. For each samples, fifteen randomly-selected raw gnocchi were analyzed.

Cooked gnocchi were evaluated after being cooked at their OCT and cooled in an airtight container for 25 min at room temperature in order to ensure a complete cooling as temperature can affect the texture properties. Cooked samples were characterized in terms of surface color and geometrical indices (as previously reported for raw sample), weight increase, by weighing gnocchi before and after cooking, and solid loss into the cooking water, by evaluating the dry matter of the cooking water (dried at 105 °C, to constant weight). All the measurements were done in triplicate, cooking fifteen gnocchi in each replicate. Gnocchi texture was assessed with a TA-HDplus Texture Analyzer (Stable Micro Systems, Surrey, UK) equipped with a 10-blade Kramer shear cell and a 250-N load cell. The Texture Exponent TEE32 V 3.0.4.0 software (Stable Micro System, Surrey, UK) was used to control the instrument and for data acquisition. Cooked gnocchi (98 ± 10 g) were compressed, sheared, and extruded through the bottom openings of the Kramer cell by the blades moving at 2 mm/s speed, simulating chewing. The maximum force (N) reached during the shear/extrusion test was extrapolated from the stress–deformation curve as an index of the product hardness. Seven replicates were carried out for each sample.

2.4. Sensory Quality Evaluation

Ninety-six consumers (36 males and 60 females, age range: 19–64 years, mean age 30.3 ± 11.8) took part in the experiment. All subjects reported to like gnocchi and to consume them at least once in a month. Participants with allergies or intolerances towards ingredients present in the formulations were excluded from the evaluation. All subjects gave their written informed consent prior to the beginning of the study and they were instructed to refrain from smoking, eating, and drinking (except water), in the hour before tasting. The study protocol was approved by the Ethical Committee of the University of Milan. The study was conducted in agreement with the Italian ethical requirements on research activities and personal data protection (D.L. 30.6.03 n. 196) and according to the principles of the Declaration of Helsinki.

The tasting sessions were organized over two consecutive days in the sensory laboratory of the Department of Food, Environmental, and Nutritional Sciences (University of Milan, Milan, Italy) designed according to the International Organization for Standardization (ISO) guidelines [22]. Eight rounds were organized in the 11:00 a.m.–2:00 p.m. time slot, each comprising 12 subjects. For each round, the samples (300 g) were cooked in 3 L of salted water (salt: 10 g/L) and tasted with tomato sauce (40 g/100 g; provided by Zini Prodotti Alimentari S.p.A., Cesano Boscone, Italy) in order to make the evaluation more similar to common consumption. Samples were prepared immediately before each tasting session and served monadically (approximately 20 g by sample) in white plastic dishes coded with 3-digit numbers. The presentation order was balanced according to Latin square to limit carry-over effects [23].

Two methods were applied: a hedonic test to have insights on the overall liking of each sample and the Check-All-That-Apply (CATA) method, which is a simple approach to gather information about consumers' perception of the sensory characteristics of food products [24]. With this method, consumers are asked to taste the products and to answer a CATA question by selecting from a list of descriptors all the terms that they consider appropriate to describe each of the samples. In the present study, the descriptors were chosen in a preliminary test involving 6 untrained subjects who tasted the six samples and generated a list of 12 terms: 6 for texture in mouth (firm, coarse, rubbery, soft to be chewed, grainy/pieces, adhesive/sticks to teeth) and 6 hedonic terms related to appearance (pleasant and unpleasant appearance), taste (pleasant and unpleasant taste) and texture (pleasant and unpleasant texture). The terms were selected in order to be easily understood by consumers. The number of terms chosen is in line with the number suggested by the literature, i.e., between 10 and 40 terms [25].

Prior to the beginning of the session, participants were instructed about the overall methodology and received a brief explanation of the CATA terms. Then, they were invited to taste the samples in individual sensory booths under normal light conditions. Each consumer was informed about what he/she was going to taste (e.g., "You are going to taste a sample of gnocchi added with buckwheat"). For each sample, at first participants rated their overall liking using an unstructured linear scale anchored at the extremes with "Extremely disliked" (left of the scale, corresponding to 0) and "Extremely liked" (right of the scale, corresponding to 100), then, they were asked to select all the descriptors suitable for describing that sample. The position of CATA attributes in the list was randomized across participants but fixed for each participant [26]. Participants were instructed to drink a sip of water between samples tasting.

2.5. Data Analysis

Nutritional and technological results were expressed as mean ± standard deviation (SD) values. All data were subjected to one-way analysis of variance (ANOVA), followed by the Least Significant Difference (LSD) test to identify significant differences between the samples ($p \leq 0.05$). The statistical analysis was carried out using STATGRAPHIC_Plus for Windows v. 5.1 (StatPoint Inc., The Palins, VA, USA).

Sensory data were processed by a mixed ANOVA model performed on liking data considering subjects as a random factor and samples as fixed factor. The LSD test was used to compare the samples. For the CATA questions, the frequency of mention for each term was determined by counting the number of subjects that used that term to describe each sample. Cochran's Q test was performed for each of the 12 terms to evaluate significant differences among samples.

To study the relationship between CATA questions, technological properties of cooked gnocchi and liking data, Partial Least Square Regression (PLSR) analysis was performed [27,28]. PLSR models both the X- and Y-matrices simultaneously to find the variables in X that best predict the variables in Y. The PLSR components are referred to as factors or latent variables or latent structures. In PLSR models, scores and loadings express how the samples and variables are projected along the model factors [29]. CATA questions and technological variables were considered as the X matrix and average liking scores of each product as the Y matrix. Data were standardized (i.e., scaled to unit variance) prior to modeling and full cross validation was chosen as the validation method. A correlation loadings plot was used to find variables with less than 50% explained variance which were left out of the model [30]. This only resulted in the omission of one technological variable, i.e., solid loss.

SAS software v. 9.4 (SAS Institute Inc., 2012, Cary, NC, USA) and The Unscrambler X v. 10.3 (CAMO, Oslo, Norway) were used as statistical software packages. A p-value ≤ 0.05 was chosen as the threshold for statistical significance.

3. Results and Discussion

3.1. Nutritional Quality

Proximate composition of gnocchi is reported in Table 1. The amount of water added to the formulation is a critical parameter affecting the quality of the final product and approximately 53 g/100 g of water (corresponding to 49–54 g/100 g of product moisture) was indicated as a good amount for a formulation based on corn flour, rice flour, and dried potatoes [6]. Accordingly, the moisture content of GF and C was 54.7 and 53.3 g/100 g, respectively, whereas gnocchi containing R or B generally showed a higher moisture content (up to 58 g/100 g) as they required more water during pasta production to ensure a good dough workability through industrial machines, this may be due to the presence of whole meal flours having a higher fiber content. Similar moisture values (58 and 61 g/100 g) were found in the literature for gnocchi containing amaranth and quinoa flours, respectively [7].

Table 1. Nutritional composition (g/100 g) and antioxidant capacity (mmol TEAC/kg) of raw gnocchi.

	GF	GFR	GFB	C	CR	CB
Moisture	54.7 ± 0.2 bc	58.1 ± 0.2 d	56.6 ± 0.6 d	53.3 ± 0.1 b	51.1 ± 0.9 a	54.7 ± 0.9 c
Ash	1.31 ± 0.04 a	1.35 ± 0.19 a	1.38 ± 0.14 ab	1.14 ± 0.04 a	1.68 ± 0.05 c	1.67 ± 0.14 bc
Lipids	0.26 ± 0.01 a	0.47 ± 0.06 b	0.55 ± 0.04 b	0.83 ± 0.06 c	1.40 ± 0.10 e	1.15 ± 0.07 d
Proteins	4.2 ± 0.4 a	4.4 ± 0.2 a	4.6 ± 0.3 ab	6.4 ± 0.1 cd	5.5 ± 0.3 bc	6.6 ± 0.8 d
Starch *	40.3 ± 0.1 e	36.6 ± 0.2 d	31.2 ± 0.1 b	33.9 ± 0.4 c	34.5 ± 0.7 c	29.5 ± 0.1 a
Sugars	0.26 ± 0.01 a	0.34 ± 0.02 a	0.36 ± 0.08 a	4.70 ± 0.49 c	2.77 ± 0.24 b	2.66 ± 0.25 b
TDF	0.8 ± 0.3 a	1.3 ± 0.3 ab	6.0 ± 0.5 d	1.7 ± 0.1 b	1.8 ± 0.1 b	4.8 ± 0.1 c
IDF	0.6 ± 0.1 a	1.0 ± 0.3 a	5.4 ± 0.4 c	1.0 ± 0.2 a	1.0 ± 0.2 a	4.2 ± 0.2 b
SDF	0.2 ± 0.1 a	0.3 ± 0.1 a	0.7 ± 0.1 b	0.7 ± 0.1 b	0.8 ± 0.1 b	0.6 ± 0.1 b
TEAC	2.13 ± 0.18 a	3.46 ± 0.11 ab	6.93 ± 1.22 c	4.16 ± 0.66 b	4.83 ± 0.37 b	6.94 ± 0.09 c

* Starch was calculated as difference; TDF, total dietary fiber; IDF, insoluble dietary fiber; SDF, soluble dietary fiber; TEAC, Trolox equivalent antioxidant capacity. In the same row, data having different letters are significantly different ($p < 0.05$).

As expected, for both conventional and GF products the addition of wholemeal R and B flours tended to increase the levels of some nutrients (i.e., lipids, protein, and dietary fiber) while decreasing the starch content. In particular, the addition of B and R significantly increased fat content to about twice the reference samples, due to the lipids presents in the two flours used. In conventional gnocchi (sample C), the addition of both flours to the formulation resulted in a significant reduction (about 50%) of the soluble sugar content. Only the enrichment with B flour led to a significant increase in the level of total fiber in both conventional and gluten free products, mainly represented by the insoluble fraction. It is worth mentioning that the total fiber contents achieved in GFB and CB (>3%) allows one to report on the label the nutritional claim—"Source of fiber"—in accordance with the European Regulation 1924/2006 about nutrition and health claims provided on food products. The addition of whole buckwheat, a pseudocereal known to possess a high antioxidant potential [19], increased the antioxidant capacity of GFB and CB, probably promoting their protection from oxidation during storage.

Regarding the starch quality data shown in Figure 2 evidence that the addition of B and R flours led to changes in the rate of starch digestion in comparison to reference products (C and GF). In particular, the addition of red rice flour in conventional products did not exert significant effects, while in GF gnocchi it increased by about 20% the fraction of rapidly digestible starch and reduced to about a third the fraction of slowly digestible starch, thus suggesting a potential increase in post-prandial glycemia of the enriched products compared to the GF reference. In fact, the RDS fraction is directly related to the glycemic response of the product itself [20,31], while the insulinemic response appears inversely related to the SDS fraction [32]. In addition, recently, the European Food Safety Authority (EFSA) has approved a health claim regarding the role of SDS in the control of post-prandial blood glucose [33]. The different effect of R addition in the two types of reference products could be attributable, at least in part, to the different impact of rice starch on the structure: in GF the presence of rice starch weakened the structure of the finished product [34], while in traditional gnocchi, where the structure is maintained by the presence of gluten, the "destructuring" effect of rice starch was not detected. Differently, the use of buckwheat flour, affected the rate of starch digestion in both types of gnocchi, promoting a greater presence of starch that is rapidly digestible and a reduced level of the slowly digestible, more evident in GF products (RDS +16%; SDS −15%) compared to conventional ones (RDS +10%; SDS −11%), thus likely increasing the glycemic impact of both products. This effect is probably attributable to the presence of a higher fiber content (mostly insoluble), which could interfere with the formation of a compact gluten matrix [34] thus, promoting a greater accessibility of starch to digestive enzymes. In GF products, the higher RDS fraction (>starch accessibility to digestion) is likely to be due to a "destructuring" effect attributable to the presence of fiber that tied water during kneading, thus, compromising the proper distribution of water and interfering with the gelatinization of the dough starch.

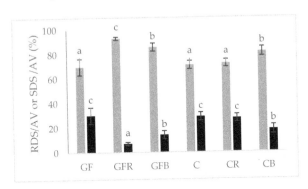

Figure 2. Rapidly (grey bar) and slowly (black bar) digestible starch fractions in cooked gnocchi. RDS, rapidly digestible starch; SDS slowly digestible starch; AV, total available starch. For each parameter, bars with different letters are significantly different ($p < 0.05$).

3.2. Technological Quality

The technological evaluation of raw and cooked gnocchi is summarized in Tables 2 and 3, respectively.

Table 2. Technological characterization of raw gnocchi.

	GF	GFR	GFB	C	CR	CB
L*	77.5 ± 2.3 [c]	58.4 ± 1.4 [b]	54.3 ± 3.2 [a]	76.2 ± 1.7 [c]	58.5 ± 2.9 [b]	54.2 ± 2.3 [a]
a*	-5.5 ± 0.5 [b]	3.5 ± 0.2 [e]	0.9 ± 0.2 [d]	-0.6 ± 0.4 [a]	3.3 ± 0.9 [e]	0.5 ± 0.3 [c]
b*	22.7 ± 2.3 [d]	7.1 ± 0.6 [b]	5.8 ± 1.0 [a]	25.3 ± 1.6 [e]	10.0 ± 0.8 [c]	6.7 ± 0.8 [ab]
Area (mm²)	447.0 ± 46.3 [a]	539.4 ± 62.6 [c]	510.6 ± 66.6 [bc]	442.4 ± 36.5 [a]	491.7 ± 63.4 [b]	534.2 ± 69.1 [c]
Width (mm)	18.7 ± 1.1 [a]	22.0 ± 1.9 [cd]	22.7 ± 2.6 [d]	19.0 ± 1.0 [a]	20.1 ± 1.8 [b]	21.6 ± 1.6 [c]
Length (mm)	30.8 ± 2.5 [ab]	32.1 ± 2.7 [bc]	30.1 ± 2.8 [a]	29.9 ± 2.1 [a]	31.8 ± 2.7 [bc]	32.7 ± 3.3 [c]

In the same row, data having different letters are significantly different ($p < 0.05$).

Table 3. Technological characterization of cooked gnocchi.

	GF	GFR	GFB	C	CR	CB
L*	69.6 ± 0.9 [e]	52.2 ± 0.8 [c]	44.7 ± 1.5 [a]	64.3 ± 1.0 [d]	51.7 ± 1.3 [c]	47.1 ± 1.6 [b]
a*	-6.7 ± 0.5 [a]	4.4 ± 0.5 [d]	1.5 ± 0.3 [c]	-6.3 ± 0.3 [b]	10.0 ± 0.5 [d]	6.9 ± 0.6 [b]
b*	21.7 ± 1.8 [f]	8.1 ± 0.7 [c]	5.1 ± 1.1 [a]	20.2 ± 1.3 [e]	14.1 ± 1 [c]	16.4 ± 1 [d]
Weight increase (g/100 g)	12.1 ± 1 [b]	14.2 ± 1 [c]	11.4 ± 1 [a]	11.5 ± 1 [ab]		
Solid loss (g/100 g)	4.06 ± 0.51 [a]	5.38 ± 1.23 [b]	3.52 ± 0.70 [a]	3.70 ± 0.60 [a]	3.93 ± 0.61 [a]	3.48 ± 0.73 [a]
Hardness (N)	408 ± 13 [f]	108 ± 2 [a]	243 ± 8 [c]	336 ± 17 [e]	200 ± 18 [b]	262 ± 8 [d]

In the same row, data having different letters are significantly different ($p < 0.05$).

The color of food is one of the first aspects noticed by the consumer and it can affect its acceptability [35]. Besides the production process, that in the present study was kept constant for all the newly-made gnocchi, pasta color is mostly due to the ingredient used (e.g., type of flour and degree of milling). Color of unconventional flours (R and B) was evaluated: As expected, R flour was characterized by high redness and yellowness values ($a^* = 4.7 \pm 0.2$; $b^* = 10.3 \pm 0.3$), whereas B had a white-yellow color (a^* value around zero; $b^* = 5.7 \pm 0.3$). Lightness values (L*) were 70.6 ± 0.5 and 74.8 ± 1.4 for R and B flour, respectively. Consequently, as reported in Table 2, GFR and CR gnocchi were significantly redder than the other samples that had a* values ranging from −5.5 to 0.9. The presence of buckwheat (i.e., GFB and CB samples) caused a reduction of lightness in comparison to the conventional sample and intermediate a* and b* values. Tiny differences were noticed between gluten free and conventional samples characterized by high values of lightness (L* = 76–77) and yellowness (b* = 23–25) values. Similar L* values (77.5 and 63.39–78.34) were reported for reference gnocchi [7,36]. Yellow color in fresh pasta is generally considered an important quality attribute and both GF and C gnocchi showed values in agreement with literature data (b* = 18.5–24 [35] and b* = 22.3 [7]).

As Italian gnocchi are recognized by the consumers for their unique shape (Figure 1), the geometrical indices of raw samples were detected. Gnocchi showed an area ranging from 447 to 534 mm^2, a width of 18.7–22.7 mm, and a length of 29.9–32.7 mm. No significant differences were noticed between conventional and gluten free references suggesting that ingredients and process conditions were appropriate also in absence of gluten. The addition of wholemeal flours turned out in significant ($p < 0.05$) bigger samples probably due to the fiber swelling and the higher amount of water added that modify the viscosity of the dough and, therefore, the product shaping.

Color evaluation of cooked samples (Table 3) reflected the differences noticed in raw samples; indeed, a strong correlation (r > 0.98, $p < 0.0005$) among L*, a*, and b* values of raw and cooked samples was found, suggesting that R and B flours can be effectively used to confer particular color even to the cooked product. The addition of R and B flours led to a significantly ($p < 0.05$) higher weight increase both in gluten free and conventional samples, maybe due to the higher presence of fiber (Table 1) that binds more water during cooking. All the sample maintained their shape during cooking and they increase principally in length, as evidence in literature [1]. GFR showed also the highest area increase during cooking (3% vs. 0–2% of the other samples, data not shown). Contrary to expectations, no significant ($p < 0.05$) difference among cooking loss values were found, except for GFR that showed a higher solid loss confirming that the addition of unconventional ingredients, such as red rice flour, to a gluten free matrix is more difficult than for conventional samples. The amount of solid loss in cooking water, in fact, is widely used as an indicator of pasta quality: low amounts of residue indicate high cooking quality [1,6,35]. In general, according to literature data [37], all the enriched gnocchi showed good cooking quality having cooking loss < 6 g/100 g. In order to limit the cooking loss, a valuable strategy could be the addition of milk and eggs to the formulation as suggested in a previous study reporting cooking loss < 1 g/100 g even if quinoa and amaranth flour were added to the recipe [7]. In fact, egg proteins can ensure cohesiveness of the dough, mainly when heated [38]. Furthermore, no-forming gluten proteins can create an alternative structure preventing cooking loss [1], and proteins can interact with other compounds, such as starch or albumins, preventing starch leaching [39,40]. According to literature data [36], gnocchi is preferred to be quite thick after cooking and they should not disintegrate even if slightly overcooked. No target consistencies are known based on literature data and many different tests can be performed to investigate gnocchi texture. In the present study, both instrumental texture (Table 3) and sensory texture acceptance (Table 4) were investigated. The hardness of cooked samples evaluated by Kramer test was significantly ($p < 0.05$) affected by the different flours used: both gluten free and conventional gnocchi made with rice flour were 74 and 40% less firm than the references, whereas gnocchi made with buckwheat flour showed intermediate hardness, but in any case, lower than the references (−40% and −22% for gluten free and conventional products, respectively). Similarly, gnocchi made with quinoa and amaranth were reported to be less hard and springy than the commercial ones [7]. As previously mentioned for cooking loss, textural parameters could also be affected by the presence of fiber (which may lead to the formation of discontinuities inside pasta structure [41]), protein matrix [38], and starch organization [1]. Also flour particle size distribution could affect gnocchi texture, since literature data indicate that pasta made of fine (138–165 μm) rice flour is sticky and less hard than sample made with bigger flour [42], whereas rice flour particles < 63 µm improve noodle texture due to a rapid retrogradation of starch and consequently an increase in gel firmness [43]. R flour used in the present study had the following particle size: 1% ≤ 90, 90 < 17% ≤ 125, 125 < 33% ≤ 250, 250 < 41% ≤ 500, and 8% > 500 µm; B flour was characterized by a higher amount of small (<90 µm) and large (>500 µm) particle size: 29% ≤ 90, 90 < 16% ≤ 125, 125 < 30% ≤ 250, 250 < 11% ≤ 500, and 14% > 500 µm.

Table 4. Frequency mention (%) of Check-All-That-Apply (CATA) items for each cooked gnocchi sample.

CATA Items	Q	GF	GFR	GFB	C	CR	CB
Pleasant appearance	84.1	79.2	38.5	50.0	84.4	62.5	50.0
Unpleasant appearance	37.0	5.2	26.0	18.8	3.1	10.4	17.7
Pleasant taste	43.4	65.6	47.9	44.8	80.2	65.6	55.2
Unpleasant taste	55.1	1.0	18.8	26.0	1.0	5.2	13.5
Firm	25.2	42.7	20.8	29.2	44.8	37.5	25.0
Coarse	255.0	26.0	57.3	84.4	0.0	14.6	80.2
Rubbery to chew	45.3	53.1	19.8	28.1	49.0	49.0	25.0
Soft to chew	31.8	31.3	56.3	24.0	44.8	44.8	34.4
Grainy/pieces	182.0	9.4	44.8	63.5	1.0	18.8	71.9
Adhesive/Sticks to theet	139.9	14.6	71.9	29.2	12.5	58.3	18.8
Pleasant texture	63.9	56.3	20.8	26.0	63.5	33.3	31.3
Unpleasant texture	73.8	5.2	46.9	39.6	9.4	24.0	34.4

All values are significant at *p* < 0.0001.

3.3. Sensory Quality

Understanding the drivers of liking and disliking of GF products is important considering that food appearance, aroma, taste, and texture play a key role in food appreciation and, thus, in its consumption. In fact, dissatisfaction with both the availability and the hedonic dimension of GF products has a decisive impact on the non-compliance with gluten-free diet [44]. There was a significant difference among gnocchi samples in terms of overall acceptability (*p* < 0.0001). The multiple comparison test (Figure 3) revealed that C sample obtained the highest score but it was statistically comparable to GF sample which, in turn, was not significantly different from CR sample. The CB, GFB, and GFR samples were comparable to each other, but scored significantly lower and were significantly less appreciated than the other samples.

The frequency table of terms checked by consumers to describe the gnocchi samples is reported in Table 4. Significant differences (*p* < 0.0001 for all items) were found in the frequency mention for all CATA items. The most appreciated samples (C and GF) were associated more frequently with descriptors such as pleasant appearance, taste, and texture. On the other hand, characteristics such as grainy and coarse texture were used to describe samples with integration of buckwheat, while the integration of rice gave samples a sticky texture, especially to sample GFR (71.9%), which was also perceived as softer (56.3%) than the other samples.

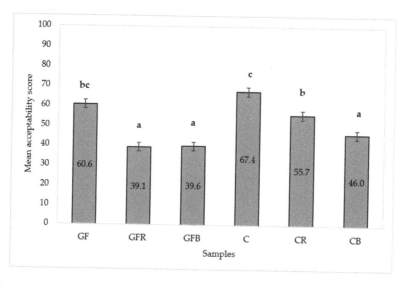

Figure 3. Mean acceptability scores with standard error of the mean. Different letters indicate significant differences according to Least Significant Difference (LSD) post-hoc test.

Scores and loadings plots from PLSR performed on technological variables, CATA items and liking are reported in Figure 4a,b. The purpose of this calculation was to establish which technological variables and sensory attributes predict the preference for the samples. The first factor explains respectively the 71% and 95% of the variation in Y, while the second factor accounts for respectively the 14% and 4% of variation. The Y variable (LIKING) is located in the upper right quadrant (Figure 4b). As the first factor explains almost all the information in the model, variables having a positive coordinate on the first factor show a direct correlation with preference, while variables with a negative coordinate on factor 1 are negatively correlated to preference. Texture was a major contributor to liking and rejection of the samples. Firm and rubbery texture properties were positive predictors of liking, whereas a granular and coarse matrix contributed negatively to liking. Firmness perceived by consumers was positively correlated with maximum force. L* and b* colorimetric coordinates also contributed positively to liking, while a* contributed negatively (Figure 4b). Comparing the scores (Figure 4a) and loadings (Figure 4b) plots, gnocchi without addition (C and GF) were the most preferred, because they were characterized by higher firmness and by bright/yellow color (as expressed by L* and b* parameters). Samples with rice addition (GFR and CR) had a soft and adhesive texture, which was related to a higher weight increase after cooking, and a higher intensity of red color as expressed by the a* parameter. These properties were disliked by consumers. GF products are often reported being of poorer sensory quality compared to conventional products, especially with regards to texture due to the lack of viscoelastic properties imparted by gluten. GF pasta is generally characterized by high stickiness, low firmness, and is prone to important cooking loss [1,45]. Hydrocolloids and gums have been reported to improve firmness and mouthfeel sensations of GF pasta formulations because they are able to create a network which contributes to a better perceived texture [46]. Samples with buckwheat addition (GFB and CB) were also rejected by consumers due to their grainy and coarse texture likely due to the higher amounts of particles having dimensions higher than 500 μm and to a higher fiber content, which also imparted an unpleasant taste and appearance to these samples. Previous studies on wheat bran enriched pasta showed that fibers elicit negative sensations in consumers such as dark color, bitter taste, and a coarse texture, which can make food unpalatable [47]. Consumers like to be in full control of the food placed in their mouth. In this context, food containing unexpected lumps or hard particles are usually rejected for fear of gagging or choking [48]. GF products lack of many important nutrients including dietary fibers because they are usually obtained from refined flour and/or starches that are not enriched or fortified [46]. Buckwheat is a valuable source of fiber, therefore, its incorporation in GF formulations is important and should be optimized. The present findings indicate that texture properties of both conventional and GF formulations added with buckwheat should be improved by reducing particles size thus making the matrix more uniform and palatable.

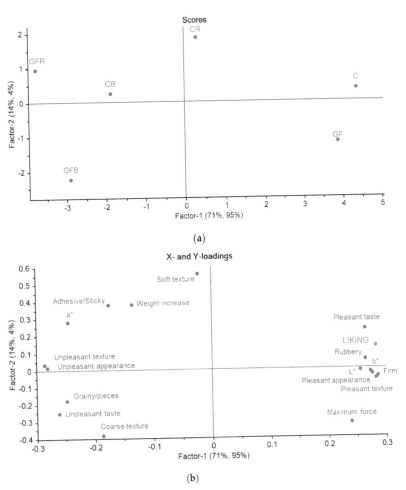

Figure 4. Scores (**a**) and loadings (**b**) plots obtained by the Partial Least Square Regression (PLSR) model of the six gnocchi samples based on CATA questions, technological variables, and liking.

4. Conclusions

The addition of 20% of rice and buckwheat whole meal flour both to conventional and gluten free gnocchi caused important changes in the nutritional, technological, and sensory properties of the cooked products. Rice flour addition highly modified GF pasta structure, increasing solid loss in cooking water and reducing product firmness, while the same flour added to conventional gnocchi determined a limited increase of cooking losses and a lower softening after cooking and a higher appreciation, though an excessive stickiness represented the main defect.

Foods **2021**, 10, 91

Buckwheat addition has allowed for a fiber content in both conventional and GF gnocchi higher than 3% and, thus, the possibility to report on the label the nutritional claim—"Source of fiber"—in accordance with the European Regulation 1924/2006. Furthermore, cooking losses were not affected by buckwheat addition indicating a better texturizing capacity in comparison to rice. On the other hand, great attention has to be paid to the particle size of wholemeal buckwheat as a high number of particles greater than 500 μm (mainly derived from the seed coat grinding) produces negative sensations in the consumers, thus reducing the product acceptability.

Author Contributions: Conceptualization, C.A., M.L. (Mara Lucisano), and M.V.; Data curation and methodology, C.C., M.C.C., and M.L. (Monica Laureati); Formal analysis, C.C., M.C.C., D.E., and M.L. (Monica Laureati); Investigation, C.C., M.C.C., D.E., and M.L. (Monica Laureati); Project administration, M.L. (Mara Lucisano) and M.V.; Resources, M.L. (Mara Lucisano) and M.V.; Supervision, C.A. and M.L. (Mara Lucisano); Writing—original draft C.A., C.C., M.C.C., D.E., and M.L. (Monica Laureati); Writing—review and editing, M.L. (Mara Lucisano) and M.V. All authors have read and agreed to the published version of the manuscript.

Funding: This research was supported by Lombardy Region (Linea R&S per Aggregazioni; project number 145075).

Institutional Review Board Statement: The study was conducted according to the guidelines of the Declaration of Helsinki, and approved by the Ethics Committee of the University of Milan (protocol code 32/12, date of approval 16 November 2012).

Informed Consent Statement: Informed consent was obtained from all subjects involved in the study.

Acknowledgments: We would like to thank the Department of Food, Environmental and Nutritional Sciences, Università degli Studi di Milano, for partially covering the open access APC.

Conflicts of Interest: The authors declare no conflict of interest.

References

1. Lucisano, M.; Cappa, C.; Fongaro, L.; Mariotti, M. Characterisation of gluten-free pasta through conventional and innovative methods: Evaluation of the cooking behaviour. *J. Cereal Sci.* **2012**, 56, 667–675. [CrossRef]
2. Mariotti, M.; Iametti, S.; Cappa, C.; Rasmussen, P.; Lucisano, M. Characterisation of gluten-free pasta through conventional and innovative methods: Evaluation of the uncooked products. *J. Cereal Sci.* **2011**, 53, 319–327. [CrossRef]
3. Carini, E.; Curti, E.; Minucciani, M.; Antoniazzi, F.; Vittadini, E. Pasta. In *Engineering Aspects of Cereal and Cereal-Based Products*; Guiné, R.P.F., Correia, P.M.R., Eds.; CRC Press: Boca Raton, FL, USA, 2014; Chapter 10; pp. 211–238. [CrossRef]
4. Giuberti, G.; Gallo, A. Reducing the glycaemic index and increasing the slowly digestible starch content in gluten-free cereal-based foods: A review. *Int. J. Food Sci. Technol.* **2018**, 53, 50–60. [CrossRef]
5. Friedel, J.; Glattes, H.; Schleining, G. Austrian dumplings. In *Traditional Foods. General and Consumer Aspects*; Kristbergsson, K., Oliveira, J., Eds.; Springer: New York, NY, USA, 2016; Chapter 10; pp. 139–156.
6. Cappa, C.; Franchi, R.; Bogo, V.; Lucisano, M. Cooking behavior of frozen gluten-free potato-based pasta (gnocchi) obtained through turbo cooking technology. *LWT Food Sci. Technol.* **2017**, 84, 464–470. [CrossRef]
7. Burgos, V.E.; López, E.P.; Goldner, M.C.; Del Castillo, V.C. Physicochemical characterization and consumer response to new Andean ingredients-based fresh pasta: Gnocchi. *Int. J. Gastron. Food Sci.* **2019**, 16, 100142. [CrossRef]
8. Budryn, G.; Nebesny, E.; Rachwał-Rosiak, D.; Oracz, J. Fatty acids, essential amino acids, and chlorogenic acids profiles, in vitro protein digestibility and antioxidant activity of food products containing green coffee extract. *Int. Food Res. J.* **2013**, 20, 2133–2144.
9. Liu, T.; Hamid, N.; Yoo, M.J.Y.; Kantono, K.; Pereira, L.; Farouk, M.M.; Knowles, S.O. Physicochemical and sensory characterization of gnocchi and the effects of novel formulation on in vitro digestibility. *J. Food Sci. Technol.* **2016**, 53, 4033–4042. [CrossRef]
10. Lang, G.H.; Rockenbach, B.A.; Ferreira, C.D.; de Oliveira, M. Delayed drying interval of red rice: Effects on cooking properties, in vitro starch digestibility and phenolics content. *J. Stor. Prod. Res.* **2020**, 87, 101613. [CrossRef]
11. Mbanjo, E.G.N.; Kretzschmar, T.; Jones, H.; Ereful, N.; Blanchard, C.; Boyd, L.A.; Sreenivasulu, N. The genetic basis and nutritional benefits of pigmented rice grain. *Front. Genet.* **2020**, 11, 229. [CrossRef]
12. Kasunmala, I.G.G.; Navaratne, S.B.; Wickramasinghe, I. Effect of process modifications and binding materials on textural properties of rice noodles. *Int. J. Gastron. Food Sci.* **2020**, 21, 100217. [CrossRef]
13. Manaois, R.V.; Zapater, J.E.I.; Labargan, E.S.A. Nutritional qualities, antioxidant properties and sensory acceptability of fresh wheat noodles formulated with rice bran. *Int. Food Res. J.* **2020**, 27, 308–315.
14. Janssen, F.; Pauly, A.; Rombouts, I.; Jansens, K.J.A.; Deleu, L.J.; Delcour, J.A. Proteins of amaranth (*Amaranthus* spp.), buckwheat (*Fagopyrum* spp.), and quinoa (*Chenopodium* spp.): A food science and technology perspective. *Compr. Rev. Food Sci. Food Saf.* **2017**, 16, 39–58. [CrossRef] [PubMed]

15. Martínez-Villaluenga, C.; Peñas, E.; Hernández-Ledesma, B. Pseudocereal grains: Nutritional value, health benefits and current applications for the development of gluten-free foods. *Food Chem. Toxicol.* **2020**, *137*, 111178. [CrossRef] [PubMed]

16. AACC (American Association of Cereal Chemists). *Approved Methods of the AACC; n. 44-15.02; 08-01.01; 30-10.01; 46-12.01; 32-07.01*, 10th ed.; AACC: St Paul, MN, USA, 2000.

17. Rocklin, R.D.; Pohl, C.A. Determination of carbohydrates by anion exchange chromatography with pulsed amperometric detection. *J. Liq. Chromatogr.* **1983**, *6*, 1577–1590. [CrossRef]

18. AOAC (Association of Official Analytical Chemists). *Official Methods of Analysis*, 16th ed.; AOAC: Gaithersburg, MD, USA, 1999.

19. Serpen, A.; Gökmen, V.; Pellegrini, N.; Fogliano, V. Direct measurement of the total antioxidant capacity of cereal products. *J. Cereal Sci.* **2008**, *48*, 816–820. [CrossRef]

20. Englyst, K.N.; Englyst, H.N. Rapidly available glucose in foods: An in vitro measurement that reflects the glycemic response. *Am. J. Clin. Nutr.* **1999**, *69*, 448–454. [CrossRef]

21. Englyst, K.N.; Hudson, G.J. Starch analysis in Food. *Encycl. Anal. Chem.* **2000**, *66*, 4262–4264.

22. ISO (International Organization for Standardization) 8589:2007. *Sensory Analysis—General Guidance for the Design of Test Rooms*; ISO: Geneva, Switzerland, 2007.

23. Macfie, H.J.; Bratchell, N.K. Designs to balance the effect of order of presentation and first-order carry-over effects in hall tests. *J. Sens. Sci.* **1989**, *4*, 129–148. [CrossRef]

24. Varela, P.; Ares, G. Sensory profiling, the blurred line between sensory and consumer science. A review of novel methods for product characterization. *Food Res. Int.* **2012**, *48*, 893–908. [CrossRef]

25. Ares, G.; Jaeger, S.R. Examination of sensory product characterization bias when check-all-that-apply (CATA) questions are used concurrently with hedonic assessments. *Food Qual. Prefer.* **2015**, *40*, 199–208. [CrossRef]

26. Meyners, M.; Castura, J.C. Randomization of CATA attributes: Should attribute lists be allocated to assessors or to samples? *Food Qual. Prefer.* **2016**, *48*, 210–215. [CrossRef]

27. Laureati, M.; Cattaneo, C. Application of the check-all-that-apply method (CATA) to get insights on children's drivers of liking of fiber-enriched apple purees. *J. Sens.* **2017**, *32*, e12253. [CrossRef]

28. Meyners, M.; Castura, J.C.; Carr, B.T. Existing and new approaches for the analysis of CATA data. *Food Qual. Prefer.* **2013**, *30*, 309–319. [CrossRef]

29. Wold, S.; Sjöström, M. PLS-regression: A basic tool of chemometrics. *Chemometr. Intell. Lab.* **2001**, *58*, 109–130. [CrossRef]

30. Martens, H.; Martens, M. Modified Jack-knife estimation of parameter uncertainty in bilinear modelling by partial least squares regression. *Food Qual. Prefer.* **2000**, *11*, 5–16. [CrossRef]

31. Englyst, H.N.; Veenstra, J. Measurement of rapidly available glucose (RAG) in plant foods: A potential in vitro predictor of the glycemic response. *Br. J. Nutr.* **1996**, *75*, 327–337. [CrossRef] [PubMed]

32. Garsetti, M. The glycemic and insulinemic index of plain sweet biscuits: Relationships to in vitro starch digestibility. *J. Am. Coll. Nutr.* **2005**, *24*, 441–447. [CrossRef] [PubMed]

33. EFSA Panel on Dietetic Products; Nutrition and Allergies (NDA). Scientific Opinion on the substantiation of a health claim related to "slowly digestible starch in starch-containing foods" and "reduction of post-prandial glycaemic responses" pursuant to Article 13 of Regulation (EC) No 1924/20061. *EFSA J.* **2011**, *9*, 2. [CrossRef]

34. Marti, A.; Abbasi Parizad, P.; Marengo, M.; Erba, D.; Pagani, M.A.; Casiraghi, M.C. In Vitro Starch Digestibility of Commercial Gluten-Free Pasta: The Role of Ingredients and Origin. *J. Food Sci.* **2017**, *82*, 1012–1019. [CrossRef]

35. Carini, E.; Vittadini, E.; Curti, E.; Antoniazzi, F. Effects of different shaping modes on physico-chemical properties and water status of fresh pasta. *J. Food Eng.* **2009**, *93*, 400–406. [CrossRef]

36. Alessandrini, L.; Balestra, F.; Romani, S.; Rocculi, P.; Rosa, M.D. Physicochemical and sensory properties of fresh potato-based pasta (Gnocchi). *J. Food Sci.* **2010**, *75*, S542–S547. [CrossRef] [PubMed]

37. Hummel, C. *Macaroni Products: Manufacture, Processing and Packing*; Food Trade Press: London, UK, 1966.

38. Alamprese, C.; Casiraghi, E.; Pagani, M.A. Development of gluten-free fresh egg pasta analogues containing buckwheat. *Eur. Food Res. Technol.* **2007**, *225*, 205–213. [CrossRef]

39. Alamprese, C.; Iametti, S.; Rossi, M.; Bergonzi, D. Role of pasteurisation heat treatments on rheological and protein structural characteristics of fresh egg pasta. *Eur. Food Res. Technol.* **2005**, *221*, 759. [CrossRef]

40. Fiorda, F.A.; Soares, M.S., Jr.; da Silva, F.A.; Grosmannb, M.V.E.; Souto, L.R.F. Amaranth flour, cassava starch and cassava bagasse in the production of gluten-free pasta: Technological and sensory aspects. *Int. J. Food Sci. Technol.* **2013**, *48*, 1977–1984. [CrossRef]

41. Petitot, M.; Boyer, L.; Minier, C.; Micard, V. Fortification of pasta with split pea and faba bean flours: Pasta processing and quality evaluation. *Food Res. Int.* **2010**, *43*, 634–641. [CrossRef]

42. Hemavathy, J.; Baht, K.K. Effect of particicle size on viscoamylographic behavior of rice flour and vermicelli quality. *J. Texture Stud.* **1994**, *25*, 469–476. [CrossRef]

43. Nura, M.; Kharidah, M.; Jamilah, B.; Roselina, K. Textural properties of laksa noodle as affected by rice flour particle size. *Int. Food Res. J.* **2011**, *18*, 1309–1312.

44. Laureati, M.; Giussani, B.; Pagliarini, E. Sensory and hedonic perception of gluten-free bread: Comparison between celiac and non-celiac subjects. *Food Res. Int.* **2012**, *46*, 326–333. [CrossRef]

45. Woomer, J.S.; Akinbode, A.A. Current applications of gluten-free grains—A review. *Crit. Rev. Sci. Nutr.* **2021**, *61*, 14–24. [CrossRef]

46. Padalino, L.; Conte, A.; Del Nobile, M.A. Overview on the General Approaches to Improve Gluten-Free Pasta and Bread. *Foods* **2016**, *5*, 87. [CrossRef]

47. Laureati, M.; Conte, A. Effect of fiber information on consumer's expectation and liking of wheat bran enriched pasta. *J. Sens. Stud.* **2016**, *31*, 348–359. [CrossRef]

48. Szczesniak, A.S. Texture is a sensory property. *Food Qual. Prefer.* **2002**, *13*, 215–225. [CrossRef]

foods

MDPI

Article

Wheat Bread Fortification by Grape Pomace Powder: Nutritional, Technological, Antioxidant, and Sensory Properties

Roberta Tolve [1], Barbara Simonato [1,*], Giada Rainero [1], Federico Bianchi [1], Corrado Rizzi [1], Mariasole Cervini [2] and Gianluca Giuberti [2]

1   Department of Biotechnology, University of Verona, Strada Le Grazie 15, 37134 Verona, Italy; roberta.tolve@univr.it (R.T.); giada.rainero@univr.it (G.R.); federico.bianchi_02@univr.it (F.B.); corrado.rizzi@univr.it (C.R.)
2   Department for Sustainable Food Process, Università Cattolica del Sacro Cuore, via Emilia Parmense 84, 29122 Piacenza, Italy; mariasole.cervini@univr.it (M.C.); gianluca.giuberti@unicatt.it (G.G.)
*   Correspondence: barbara.simonato@univr.it

Abstract: Grape pomace powder (GPP), a by-product from the winemaking process, was used to substitute flour for wheat bread fortification within 0, 5, and 10 g/100 g. Rheological properties of control and fortified doughs, along with physicochemical and nutritional characteristics, antioxidant activity, and the sensory analysis of the obtained bread were considered. The GPP addition influenced the doughs' rheological properties by generating more tenacious and less extensible products. Concerning bread, pH values and volume of fortified products decreased as the GPP inclusion level increased in the recipe. Total phenolic compounds and the antioxidant capacity of bread samples, evaluated by FRAP (ferric reducing ability of plasma) and ABTS (2,2'-azino-bis (3-ethylbenzothiazoline-6-sulfonic acid)) assays, increased with GPP addition. Moreover, the GPP inclusion level raised the total dietary fiber content of bread. Regarding sensory evaluation, GPP fortification had a major impact on the acidity, the global flavor, the astringency, and the wine smell of bread samples without affecting the overall bread acceptability. The current results suggest that GPP could be an attractive ingredient used to obtain fortified bread, as it is a source of fiber and polyphenols with potentially positive effects on human health.

Keywords: bread fortification; grape pomace; agro-industrial by-products; antioxidant activity; phenolic compounds; sensory analysis

check for updates

Citation: Tolve, R.; Simonato, B.; Rainero, G.; Bianchi, F.; Rizzi, C.; Cervini, M.; Giuberti, G. Wheat Bread Fortification by Grape Pomace Powder: Nutritional, Technological, Antioxidant, and Sensory Properties. *Foods* **2021**, *10*, 75. https://doi.org/10.3390/foods10010075

Received: 8 December 2020
Accepted: 29 December 2020
Published: 2 January 2021

Publisher's Note: MDPI stays neutral with regard to jurisdictional claims in published maps and institutional affiliations.

1. Introduction

White wheat bread is a worldwide staple food, rich in complex carbohydrates (i.e., starch) and generally poor in dietary fiber and other micro-and macronutrients [1]. However, consumer demand for healthy and high-nutritional-value foods is increasing, and this phenomenon has attracted the attention of food manufacturers.

Specifically, the UN has set new challenges that must be achieved by the human population. Among the 17 extremely important goals of Agenda 2030, "Responsible consumption and production" and "Health and well-being" are reported. Particularly, the division between economic growth and environmental depletion, the enhancing of resource efficiency, and the promotion of sustainable lifestyles should be the starting point of ecofriendly consumption and production [2].

Grape pomace (GP), the main residue from the winemaking process, represents a promising by-product. It has been evaluated that 17 kg of GP is discarded for about every hectoliter of wine produced [3]. So, GP, in terms of a sustainable economy, is a putative ingredient to reduce industrial waste and to promote economic profit. In addition, GP contains several bioactive compounds, such as polyphenols and dietary fibers, known for their healthy properties [4,5]. Several epidemiological studies on human health have underlined the beneficial role of phenolic compounds in the prevention of several diseases [6–8]. In addition, a greater consumption of dietary fiber could reduce the risk associated with the incidence of certain forms of cancer and the development of diabetes. Additionally, dietary fiber improves the bowel transit of feces and the feeling of satiety, reduces blood cholesterol levels, and prevents obesity [9–12]. However, the recommended intakes (25–30 g/day) are rarely reached by consumers, and market availability of fiber-enriched or fortified foods can help them to achieve the correct daily intake [13].

From this point of view, white wheat bread could be a perfect target for GP fortification [14–19]. In this context, it is well known that the addition of new ingredients in white bread formulations generally leads to various changes in technological and nutritional properties [20–22].

In a framework of developing innovative wheat-based bread, it is therefore essential to assess both the nutritional and technological effect of emerging unconventional ingredients following breadmaking. Then, innovative fortified food products must be subjected to a sensory evaluation to verify their acceptability and to assess how the inclusion of a specific ingredient could modify the sensory profile of the final food product [23].

In a previous study, Hayta et al. [24] concluded that GP powder inclusion significantly contributed to the improvement of bread functional properties, evaluating the antiradical activity, total phenolic content, physicochemical, textural, and sensory properties of the obtained product. In fact, in a framework of developing innovative wheat-based bread, it is essential to assess both the nutritional and technological effect of emerging unconventional ingredients following breadmaking. Then, innovative fortified food products must be subjected to a sensory evaluation to verify their acceptability and to assess how the inclusion of a specific ingredient could modify the sensory profile of the final food product [23]. However, the inclusion of an ingredient rich in fiber could also strengthen the structure of bread doughs, reducing the doughs' extensibility and affecting the bread volume and texture [25]. Therefore, this study, in addition to evaluating the effect of GP inclusion on dough rheological properties, dealt with the investigation of technological, sensory, and nutritional properties of bread fortified with increasing levels of GP.

2. Materials and Methods

2.1. Ingredients and Breadmaking

Common white wheat flour was kindly supplied by Macinazione Lendinara SpA (Arcole, Italy). The wheat flour label detailed the following composition: fat 1.2 g/100 g, total carbohydrates 71 g/100 g, protein 11 g/100 g, and dietary fiber 2.3 g/100 g.

Grape pomace (Vitis vinifera cv. Corvina) was kindly provided by Cantina Ripa Della Volta (Verona, Italy). After alcoholic fermentation, GP was pressed and dried in a vacuum oven (VD 115 Binder GmbH, Tuttlingen, Germany; 40 °C, 30 kPa) until the final moisture content of 11.0 g water/100 g dry matter (DM) was reached. The dried pomaces, without grape seeds, stems, and stalks, were ground (GM200 Retsch, Haan, Germany) to a particle size of <200 μm. The GP powder (GPP) was preserved in vacuum packaging at room temperature until analyzed or used for bread preparation as described by Cisneros-Yupanqui et al. [26]. The chemical composition of the dried GPP was as follows: crude protein: 11.19 ± 0.97 g/100 g DM; total dietary fiber: 52.3 ± 2.1 g/100 g DM; ash: 4.17 ± 0.87 g/100 g DM.

Experimental recipes were obtained by replacing white wheat flour with 0, 5, and 10 g/100 g of GPP, obtaining GP0, GP5, and GP10 bread samples, respectively. The recipe was based on 320 g of composite flours, 210 mL of tap water, 3 g of salt, 15 g of sugar, and 3.5 g of dried brewer's yeast. The breads were prepared with a commercial breadmaking machine (Panexpress 750 Metal, model 0132/00—Ariete, Italy). Doughs were mixed, fermented for 39 min at 28 °C, and then baked at 170 °C for 65 min. For each formulation, three different doughs (i.e., batches replicates) were made on the same day.

2.2. Dough Characterization

A Brabender Farinograph (Brabender, Duisburg, Germany) was used to evaluate the dough mixing properties (AACC method 54-21.02) [27]. Dough water absorption, stability, development time, degree of softening (12 min after maximum), and quality number were considered.

The viscoelastic behavior of the doughs was investigated using an alveograph (Chopin Technologies, Villeneuve La Garenne, France) (AACC method 54-30) [27]. The parameters recorded were deformation energy (W), tenacity (P), dough extensibility (L), swelling index (G), and the curve configuration ratio (P/L ratio).

Start of gelatinization (°C), gelatinization maximum (AU), and gelatinization temperature (°C) were analyzed using an amylograph (Brabender, Duisburg, Germany) (AACC method 22-10) [27].

2.3. Bread Quality Characteristics

2.3.1. Water Activity, Moisture Content, Volume, Specific Volume, and Baking-Loss

The water activity (aw) of bread samples was measured with a Hygropalm HC2-AW-meter (Rotronic Italia, Milano, Italy) at 23 °C, whereas the moisture content was measured by the AACC method 44-15A [27]. The specific volume of the loaves (cm^3/g) was determined by seed displacement (AACC method 10-05.01) [27] for volume quantification (cm^3) and weight of samples (g). The baking-loss was determined as the differences in mass between the dough and the baked loaves.

2.3.2. Texture Attributes of Bread Crumb

The texture attributes in terms of firmness of bread crumb were evaluated using a texture analyzer (TVT-6700; Perten Instruments, Stockholm, Sweden) according to the AIB standard procedure for bread firmness measurement [28]. The maximum peak force of compression as bread firmness (N) was measured by a metal cylinder probe (25 mm diameter). For each treatment, five measurements were done for each batch.

2.3.3. Proximate Composition of Breads

Dry matter (DM; method 930.15), ash (method 942.05), crude protein (method 976.05), crude lipid (method 954.02 without acid hydrolysis), total starch (method 996.11, using thermostable α-amylase (Megazyme cat. no. E-BSTAA) and amyloglucosidase (Megazyme cat. no. E-AMGDF)), and total dietary fiber (method 991.43) were considered [29]. Free sugars were assessed using the Megazyme assay kit K-SUFRG 06/14 (Megazyme, Wicklow, Ireland). Batches were analyzed in triplicate.

2.3.4. Color Analysis

The color was measured by a reflectance colorimeter (illuminant D65) (Minolta Chroma meter CR-300, Osaka, Japan) based on the color system CIE – L* a* b*. Analyses were performed at five different points within the crumb and the crust area. Minolta Equations (1) and (2) were used to calculate the total color difference (ΔE):

$$\Delta E = \sqrt{\Delta L^2 + \Delta a^2 + \Delta b^2} \tag{1}$$

$$\Delta L = (L - L_0); \Delta a = (a - a_0); \Delta b = (b - b_0) \tag{2}$$

where *L*, *a*, and *b* are the measured values of the bread fortified with grape pomace, and L_0, a_0, and b_0 are the values of the control bread.

2.3.5. Determination of Total Phenolic Compounds (TPC), FRAP, and ABTS Assays

The extraction was carried out by stirring 1 g of bread sample with 15 mL of MeOH:HCl 97:3 (*v/v*) for 16 h in the dark at room temperature [30]. Supernatants were collected after centrifugation (3500× *g* for 10 min) and used for TPC, ABTS (2,2′-azino-bis (3-

ethylbenzothiazoline-6-sulfonic acid)), and FRAP (ferric reducing ability of plasma) radical scavenging activities determination. TPC determination was performed as described by Simonato et al. [31]. Two hundred microliters of extracts were mixed at room temperature with the same amount of Folin–Ciocalteau reagent. After 5 min, 4 mL of Na_2CO_3 (0.7 M) and 5.6 mL of Milli-Q water were added. The absorbance was measured at 750 nm (ATi Unicam UV2, Akribis Scientific, Cambridge, UK) after 1 h under stirring in the dark. The TPC is expressed as milligrams of gallic acid equivalent (GAE) per gram of dry matter (DM).

The FRAP solution was prepared by mixing a sodium acetate buffer (300 mM, pH 3.6), a $FeCl_3 \cdot 6H_2O$ solution (20 mM), and 10 mM of TPTZ solution in HCl 40 mM in a volume ratio of 10:1:1. Then, 1.8 mL of FRAP reagent and 1 mL of Milli-Q water were mixed with 10 µL of the methanolic extract. The absorbance was measured at 593 nm after 10 min at 37 °C [32]. Quantification is expressed as micromolars of Trolox equivalent (TE) per gram of DM.

The ABTS assay was performed starting from a stock solution of the radical cation ABTS + obtained mixing 7 mM ABTS and 2.45 mM $K_2S_2O_8$ (1:1 ratio) for 16 h at room temperature in the dark. The stock solution was brought at an absorbance of 0.72 ± 0.2 at 734 nm by dilution in Milli-Q water. ABTS + diluted solution (9.8 mL) was mixed with 0.2 mL of the methanolic extract and stirred for 30 min. Absorbance was measured at 734 nm and the results are expressed as micromolars of Trolox equivalent (TE) per gram of DM [33].

2.4. Sensory Evaluation of Breads

According to Quantitative Descriptive sensory Analysis (QDA), the sensory profile of samples was analyzed as proposed by Vilanova et al. [34]. A trained sensory panel of 16 persons (10 females and 6 males) aged between 22 and 33 years, recruited from the staff and students of the Department of Biotechnology of the University of Verona, was involved. Panelists generated 18 sensory terms and were trained to recognize their intensities. Color uniformity, porosity, crust thickness, fragrance, wine taste, yeast taste, global flavor, sweetness, saltiness, acidity, bitterness, moisture of the crumb, crust hardness, adhesiveness, grittiness, and astringency were considered as sensory attributes and evaluated using a hedonic 9-point scale, where 1 and 9 indicate the lowest and the highest intensity, respectively. For the sensory evaluation, samples were cut into 2 cm thick slices of about 10 g for each sample (including crumb and crust) and placed on a covered plate. All the coded samples were presented in a completely randomized and balanced order. Panelists also commented on the overall acceptability: mean scores above 5 were considered acceptable (neither like nor dislike).

2.5. Statistical Analysis

All data reported (i.e., mean values ± standard deviation) represent the means of at least three measurements. Analysis of variance (ANOVA), with a post hoc Tukey test at $p < 0.05$, was used for mean comparison. Statistical analyses were performed using the software XLSTAT Premium Version (2019.4.2, Addinsoft SARL, Paris, France).

3. Results and Discussion

3.1. Effect of Grape Pomace inclusion on Dough Rheological Properties

The addition of GPP significantly influenced the dough technological properties (Table 1). The amount of water absorbed by the doughs increased with the increasing levels of GPP in the recipe, ranging from 55.50% to 60.03% for GP0 and GP10 dough, respectively ($p < 0.05$). This increase is related to the higher dietary fiber content of the dough after GPP addition, as observed by Mironeasa [35]. Further, dietary fiber is characterized by a high number of hydroxyl groups which allow greater interactions with water molecules through hydrogen bonds [36]. The dough development time and its stability are suitable indicators of flour firmness, with higher values indicating a firmer dough. Compared to GP0, both GP5 and GP10 doughs showed no change in the development time (being on average

1.39 min; $p > 0.05$), while an increase in the stability was observed, ranging from 5.80 to 8.27, for GP0 and GP10, respectively ($p < 0.05$). Theoretically, the reduction in wheat gluten proteins caused by the GPP inclusion would lead to a decrease in the dough strength. In this case, the strength may have been enhanced by phenolic compounds of GPP, in line with previous findings [37]. In particular, the condensed tannins of the GPP can interact mainly with the glutenin fractions of wheat flour through hydrogen bonds and hydrophobic interactions. Due to their longer and broader conformation, condensed tannins have better access to the glutenin structure for noncovalent interactions with amino acid residues compared with the globular gliadins [38]. A higher content of phenolic substances in fortified dough samples could explain the reduction of the degree of softening [37], defined as the difference between the value recorded at the peak and the value recorded after 12 min. Moreover, quality numbers of GP5 and GP10 sharply increased compared with GP0, indicating a gain of flour ability in the production of doughs resistant to mechanical stress, as seen by Davoudi et al. [39].

Table 1. Farinograph characteristics of dough supplemented with grape pomace powder.

Sample	Water Absorption (%)	Stability (min)	Development Time (min)	Degree of Softening (UB)	Quality Number
GP0	55.50 ± 0.00 [a]	5.80 ± 0.72 [a]	1.40 ± 0.10 [a]	76.33 ± 4.73 [a,b]	41.67 ± 13.05 [a]
GP5	56.80 ± 0.17 [b]	7.50 ± 0.00 [b]	1.37 ± 0.12 [a]	81.33 ± 6.03 [a]	80.33 ± 3.06 [b]
GP10	60.03 ± 0.46 [c]	8.27 ± 0.55 [b]	1.40 ± 0.10 [a]	67.00 ± 3.00 [b]	87.33 ± 6.51 [b]

Data with different letters in each column are significantly different for $p < 0.05$.

The alveograph results are presented in Table 2. The extensibility value (L) reveals the ability of the dough to expand without breakdown, and it decreased following GPP addition. This could be explained by the high fiber content of the GPP, which could compete with the protein gluten for water absorption, forming a weaker gluten network, thus resulting in lower extensibility [40]. The tenacity of the dough (P), which indicates the gas-retaining ability of the dough, increased with the inclusion of GPP, ranging from 76.67 to 179.33 mm for GP0 and GP10, respectively, ($p < 0.05$). This can be related to the presence of stronger interactions between polysaccharides and the gluten proteins [40]. The P/L ratio is an index used for the gluten behavior. The addition of GPP significantly increased the P/L of the doughs. This could be caused by a strong interaction between cellulose contained in the fiber fraction and the flour protein, as already reported by Fendri et al. [41]. In particular, the P/L values were always above the range recommended for bakery leavened products, which should not be higher than 2 [42]. The G value (the size of bubbles after air insufflation) was significantly higher in GP0 dough than GP5 and GP10, whereas the deformation energy (W), described as the area under the curve of the alveogram, decreased in fortified dough samples due to the higher P and lower L values. Similar results were obtained in doughs containing different amounts of almond skin powder and in dough with barley husk powder [43,44].

Table 2. Alveograph characteristics of dough supplemented with grape pomace powder.

Sample	L (mm)	P (mm)	P/L	G (cm³)	W (10⁻⁴ J)
GP0	106.33 ± 11.72 [a]	76.67 ± 4.51 [a]	0.73 ± 0.12 [a]	22.97 ± 1.29 [a]	268.67 ± 8.62 [a]
GP5	30.00 ± 1.00 [b]	165.33 ± 4.62 [b]	5.52 ± 0.32 [b]	12.20 ± 0.20 [b]	222.67 ± 3.51 [b]
GP10	15.33 ± 1.15 [c]	179.33 ± 9.02 [b]	11.75 ± 1.12 [c]	8.70 ± 0.35 [c]	118.33 ± 11.72 [c]

Data with different letters in each column are significantly different for $p < 0.05$.

The incorporation of GPP significantly affected the gelatinization maximum of the dough (Table 3), defined as the maximum viscosity reached. Viscosity increased in GP5 and GP10 doughs compared with GP0 dough, as observed in doughs enriched with white grape peel flour of different particle sizes [45]. This result could be associated with polymer complexes from mixing derived from interactions between fiber and amylose and low molecular-weight amylopectin chains [46]. Moreover, starch may have been exposed to a faster gelatinization process, leading to higher water absorption from dietary fiber of GPP [35]. On the contrary, the addition of GPP had no effect on the starting temperature of starch gelatinization and the gelatinization temperature at maximum viscosity.

Table 3. Amylograph characteristics of dough supplemented with grape pomace powder.

Sample	Gelatinization Maximum (AU)	Start of Gelatinization (°C)	Gelatinization Temperature (°C)
GP0	1301.67 ± 17.56 [a]	64.00 ± 0.87 [a]	92.30 ± 0.75 [a]
GP5	1831.67 ± 42.52 [b]	64.00 ± 0.87 [a]	93.30 ± 0.52 [a]
GP10	1943.33 ± 11.55 [c]	64.50 ± 1.50 [a]	93.70 ± 0.75 [a]

Data with different letters in each column are significantly different for $p < 0.05$.

3.2. Physiochemical Characterization of Grape Pomace Powder and Breads

The moisture content, a_w, and pH values of GPP were respectively 6.17 ± 0.09 g/100 g DM, 0.33 ± 0.01, and 3.39 ± 0.01.

Technological properties and chemical values of breads in terms of water activity, moisture content, pH, volume, and specific volume along with the firmness values are reported in Table 4. The a_w, moisture content, and baking-loss of GP0, GP5, and GP10 samples showed no statistical differences, in contrast with other research work on bread fortification, where the moisture content generally increased with the degree of fortification [47–49]. The pH value decreased significantly in GPP-fortified bread samples compared with the control. Evidently, GPP inclusion lowered the pH in fortified bread samples, according to other researchers [37,50].

Table 4. Water activity, moisture content, pH, volume, and specific volume along with the firmness values and baking-loss of control bread (GP0) and bread fortified with grape pomace powder (GP5 and GP10).

Sample	Water Activity	Moisture content (%)	pH	Volume (cm³)	Specific Volume (cm³/g)	Firmness (N)	Baking-Loss (%)
GP0	0.96 ± 0.01 [a]	43.47 ± 0.42 [a]	5.73 ± 0.38 [a]	2623 ± 52 [a]	5.52 ± 0.11 [a]	1.0 ± 0.10 [a]	13.87 ± 0.00 [a]
GP5	0.97 ± 0.00 [a]	41.46 ± 1.71 [a]	4.44 ± 0.03 [b]	1691 ± 104 [b]	3.59 ± 0.27 [b]	21.8 ± 1.44 [b]	14.51 ± 1.92 [a]
GP10	0.97 ± 0.01 [a]	40.58 ± 3.70 [a]	3.98 ± 0.01 [c]	1334 ± 63 [c]	2.82 ± 0.13 [c]	23.2 ± 1.65 [c]	14.05 ± 0.77 [a]

Data with different letters in each column are significantly different for $p < 0.05$.

The leavening capacity and volume of bread depend on several factors such as the kneading condition, leavening time, rheology of the dough, type of ingredients, temperature, and type of yeast used. Inducted acidic conditions and the consequent increase in the number of positive charges in the doughs may have altered the gluten network of the samples, leading to the unfolding of gluten proteins [51]. Such structural changes would lead to a less extensible and more tenacious dough in fortified samples, as seen in alveographic and farinographic results, and finally, to a reduction of volume and specific volume (Figure 1). Moreover, acidic conditions could promote the gluten proteins' solubilization and, thus, the instability of the gluten network [52]. Indeed, endogenous proteases from wheat flour work better at a low pH and this could intensify gluten proteolysis, especially of GMP (glutenin macropolymer) [51].

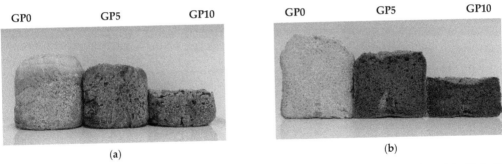

Figure 1. (**a**) Crust side of bread samples; (**b**) crumb side of bread samples. Control bread (GP0) and bread fortified with grape pomace powder (GP5 and GP10).

In particular, *Saccharomyces cerevisiae*, responsible for dough leavening, is reported to have an optimal pH growth rate ranging between 4 and 6. Nevertheless, low pH conditions could provide a more stressful environment, thus reducing the yeast activity [53]. According to Fu et al. [54] and Seczyk et al. [55], the bread volume decreased with the increase of the fortification level. In addition, the substitution of wheat flour with a high-dietary-fiber ingredient such as GPP in bread formulation could have impaired the volume of the samples. In particular, dietary fiber constitutes about 52% of GPP and can subtract water to starch granules and protein networks, affecting the overall capacity of the gluten network to retain gas bubbles [54]. High-dietary-fiber ingredients can also have a negative role on kneading by impairing air inclusion, as well as gaseous release and distribution, with a negative effect on the leavening capacity [56]. Finally, the inhibition of amylase should be also considered. Indeed, polyphenols may interact and inhibit amylase, thus contributing negatively to starch hydrolysis and to the maltose accessibility for yeasts and influencing the volume growth during the leavening stage, as suggested by other researchers [9,37].

The firmness of bread crumb increased with the increasing level of GPP, ranging from 1.0 to 23.2 N for GP0 and GP10, respectively ($p < 0.05$). These results appear consistent with those previously reported for dough and bread quality attributes. The addition of GPP may have caused a weak gluten network formation with poor gas retention ability, thus contributing to the hardening effect of bread crumb. The present findings are in agreement with Sui et al. [57], where an increase in hardness was reported in wheat-based bread crumb with increasing levels of an anthocyanin-rich black rice extract powder. Similar results have also been reported in breads formulated with increasing levels of dietary fiber extracted from culinary banana bract [58].

3.3. Chemical Composition of Breads

Among the samples, similar crude protein (being on average 12.2 g/100 g DM) and free sugar (being on average 1.8 g/100 g DM) contents were recorded (Table 5). On the contrary, the inclusion of a higher level of GPP caused a decrease in the total starch content (ranging from 85.5 to 75.3 g/100 g DM for GP0 and GP10, respectively; $p < 0.05$) and an increase in the total dietary fiber content (from 2.8 to 6.3 g/100 g DM for GP0 and GP10, respectively; $p < 0.05$). The different chemical compositions of the selected ingredients, as well as their inclusion level, can explain the present findings. Similar results have already been reported in baked foods containing GPP at different inclusion levels [50]. In addition, from a nutritional standpoint, the GP10 bread can be considered a food product high in dietary fiber, having a total dietary fiber content higher than 6 g/100 g. Higher total dietary fiber intake can prevent certain chronic diseases such as diabetes, obesity, and hypertension [59].

Table 5. Chemical composition (g/100 g dry matter) of experimental breads formulated with increasing levels of grape pomace powder (GPP) in the recipe.

Sample	Crude Lipid	Crude Protein	Total Starch	Total Dietary Fiber	Ash	Free Sugars
GP0	0.12 ± 0.05 [a]	12.4 ± 0.05 [a]	85.5 ± 2.82 [c]	2.8 ± 1.00 [a]	1.0 ± 0.01 [a]	1.6 ± 0.02 [a]
GP5	0.50 ± 0.22 [b]	12.3 ± 0.13 [a]	82.9 ± 0.89 [b]	3.9 ± 0.74 [b]	1.1 ± 0.01 [a]	1.8 ± 0.03 [a]
GP10	0.87 ± 0.06 [b]	12.1 ± 0.03 [a]	75.3 ± 1.95 [a]	6.3 ± 1.07 [c]	1.4 ± 0.02 [b]	1.9 ± 0.03 [a]

Data with different letters in each column are significantly different for $p < 0.05$.

3.4. Color Analysis

The change in color of crumb and crust of breads fortified with increasing levels of GPP is summarized in Table 6. In the control bread (GP0), the crust color was darker than the crumb due to different Maillard and caramelization reactions [60]. The GPP inclusion reversed this trend because of the GPP-darkened color parameters. According to Hayta et al. [24] and Nakov et al. [50], GPP inclusion caused a significant reduction in brightness (L*) in samples GP5 and GP10, both in the crumb and the crust. As expected, an increase in the a* parameter was evident according to the GPP increment in the bread crumb, while an opposite trend was observed in the crust. This could be due to the fact that anthocyanins, responsible for the red GPP pigmentation, are less degraded in the crumb since it undergoes lower heat treatment and maintains a higher moisture level than the crust [61]. Finally, a significant decrease in the b* parameter was observed with the progressive supplement of GPP in both crumb and crust. The total color difference (ΔE) is generally used to describe the color variation. The ΔE values revealed that GP5 and GP10 led to high color variation as the concentration of added GPP increased. This trend was observed both in the crumb and the crust but was less pronounced in the latter.

Table 6. Color analysis of bread samples expressed as L*(lightness), a* (red/green), and b* (blue/yellow) values. ΔE (total color difference)

Sample	Crust			ΔE	Crumb			ΔE
	L*	a*	b*		L*	a*	b*	
GP0	59.92 ± 1.07 [a]	6.90 ± 0.59 [a]	27.71 ± 1.13 [a]	nd	64.78 ± 1.03 [a]	1.70 ± 0.28 [a]	28.70 ± 1.41 [a]	nd
GP5	55.47 ± 0.39 [b]	5.71 ± 0.47 [b]	13.84 ± 0.28 [b]	14.61	48.34 ± 0.58 [b]	4.25 ± 0.56 [b]	17.45 ± 0.91 [b]	20.08
GP10	52.25 ± 0.65 [c]	4.62 ± 0.39 [c]	11.28 ± 0.16 [c]	18.28	47.43 ± 0.08 [b]	5.89 ± 0.06 [c]	14.74 ± 1.16 [c]	22.66

Data with different letters in each column are significantly different for $p < 0.05$. nd (not determined).

3.5. Polyphenols and Antioxidant Activity

In the present study, the total polyphenol content (TPC) and antioxidant activity of GPP and fortified bread samples were tested. GPP achieved a TPC value of 15.02 ± 0.63 mg GAE/g DM, while the antioxidant activity, assessed by FRAP and ABTS, resulted in 248.74 ± 9.53 μM TE/g DM and 213.53 ± 10.16 μM TE/g DM, respectively.

The TPC and both antioxidant tests increased significantly in fortified bread GP5 and GP10 compared with the control bread GP0, with high correlation coefficients between them (r = 0.999 CPT vs. FRAP and r = 0.979 CPT vs. ABTS) (Table 7). The TPC increased 3.5-fold and 7-fold as GPP replacement increased from 0% to 5% and from 0% to 10%, respectively. Slightly lower increases were observed by Hayta et al. [24], with 1.9-fold and 2.5-fold for the same GPP inclusion levels reported in this study. However, Hoye and Ross [62], with a substitution of 10% wheat flour by grape seed flour, reported a 20-fold increase. These discrepancies could be explained by the grape variety and the presence/absence of grapeseed flour in the dried GP [37].

Table 7. Total phenolic compounds (TPC) and antioxidant activity (FRAP and ABTS) of control bread (GP0) and bread fortified with different percentages of grape pomace (GP5 and GP10).

Sample	TPC (mg GAE 100g^{-1} DM)	FRAP (μM TE 100 g^{-1} DM)	ABTS (μM TE 100g^{-1} DM)
GP0	29.08 $\pm$ 1.45 [a]	199.72 $\pm$ 9.69 [a]	240.00 $\pm$ 7.90 [a]
GP5	101.5 $\pm$ 7.68 [b]	795.26 $\pm$ 63.11 [b]	999.50 $\pm$ 24.78 [b]
GP10	207.06 $\pm$ 9.25 [c]	1577.39 $\pm$ 87.20 [c]	1540.83 $\pm$ 47.45 [c]

Values with different superscripts within the same column are significantly different for $p < 0.01$.

3.6. Sensory Evaluation

Substitution of wheat by GPP significantly influenced most of the selected sensory attributes (Figure 2). In terms of appearance, GPP inclusion significantly affected color uniformity, porosity, and crust thickness. In particular, crust thickness decreased as the fortification level increased. As for the aroma, increasing the amount of GPP in the dough significantly decreased the overall intensity of fragrance, defined as the characteristic bread scent. Instead, as expected, the wine smell significantly increased as the GPP inclusion increased. GPP inclusion also significantly affected the taste and flavor of the bread: in particular, the global flavor and acidity increased significantly with the GPP inclusion, while the sweet taste was reduced by the GPP inclusion. In terms of texture and tactile sensations, GPP fortification significantly influenced the crumb moisture, the crust hardness, the grittiness, and the astringency. Finally, the GPP addition did not have a significant impact on the overall acceptability of the product. Indeed, the overall acceptability recorded was 6.53 $\pm$ 1.62 for GP0 bread, 6.65 $\pm$ 1.69 for GP5, and 5.59 $\pm$ 2.12 for GP10. In all the samples, the threshold value of 5 was exceeded. Likewise, Walker et al. [63] reported that bread fortified with 10% of pinot noir grape pomace was acceptable by consumers, while an inclusion of 5% was the highest fortification level for bread made with grape seed flour [62].

Figure 2. Sensory profile of quality attributes of bread fortified with different grape pomace levels (GP0 black line; GP5 gray line; GP10 light gray line). Sensory attributes were evaluated using a hedonic 9-point scale, where 1 and 9 indicate the lowest and the highest intensity, respectively.

4. Conclusions

The outcomes of our research show that GPP inclusion influences the technological, nutritional, and organoleptic properties of both dough and bread. GPP addition improved water absorption and quality number and reduced the softening degree of doughs without affecting the time and temperature of gelatinization. Moreover, GPP increased the tenacity and P/L ratio but lowered the extensibility, G value, deformation energy (W), specific volume, and pH of bread samples. GPP inclusion modified the chemical composition of bread along with the color parameters. As expected, incremental addition of GPI resulted in a significantly higher amount of TPC in bread samples and allowed increasing antioxidant activity in GP5 and GP10 compared with the control. Finally, although GPI inclusion significantly influenced aroma, taste, appearance, and flavor, a nonsignificant impact on the overall acceptability of the fortified products was observed. In conclusion GPP represents a suitable ingredient for bakery purposes since it has a significant impact on total dietary fiber as well as on the polyphenol content and antioxidant activity of the fortified bread, achieving a similar acceptability score compared to traditional bread.

Author Contributions: Conceptualization B.S.; Formal Analysis G.R., M.C., and R.T.; Methodology B.S. and G.G.; Data Curation G.G., G.R., and R.T.; Writing—Original Draft Preparation F.B., G.G., G.R., and R.T.; Writing—Review and Editing B.S., C.R., F.B., G.G., G.R., and R.T. All authors have read and agreed to the published version of the manuscript.

Funding: This research was funded by FSE 1695-0016-1463-2019 and JP 2018 (CUP B34ID18000050003)

Institutional Review Board Statement: Ethical review and approval were waived for this study since the participation was voluntary. All data were anonymous.

Informed Consent Statement: Informed consent was obtained for from all subjects involved in the study.

Data Availability Statement: Not applicable.

Acknowledgments: Common white wheat flour was kindly supplied by Macinazione Lendinara SpA as well as characterization of dough rheology (Arcole, Italy). Grape pomace (Vitis vinifera cv. Corvina) was kindly supplied by Cantina Ripa Della Volta (Verona, Italy).

Conflicts of Interest: The authors declare no conflict of interest.

References

1. Ameh, M.O.; Gernah, D.I.; Igbabul, B.D. Physico-Chemical and Sensory Evaluation of Wheat Bread Supplemented with Stabilized Undefatted Rice Bran. *Food Nutr. Sci.* **2013**, 4, 43–48. [CrossRef]
2. Transforming Our World: The 2030 Agenda for Sustainable Development. Resolution Adopted by the General Assembly on 25 September 2015. UN Doc. A/RES/70/1. Available online: https://sustainabledevelopment.un.org/post2015/transformingourworld (accessed on 20 December 2020).
3. Tolve, R.; Pasini, G.; Vignale, F.; Favati, F.; Simonato, B. Effect of Grape Pomace Addition on the Technological, Sensory, and Nutritional Properties of Durum Wheat Pasta. *Foods* **2020**, 9, 354. [CrossRef] [PubMed]
4. Padalino, L.; D'Antuono, I.; Durante, M.; Conte, A.; Cardinali, A.; Linsalata, V.; Mita, G.; Logrieco, A.F.; Del Nobile, M.A. Use of olive oil industrial by-product for pasta enrichment. *Antioxidants* **2018**, 7, 59. [CrossRef] [PubMed]
5. Simonato, B.; Trevisan, S.; Tolve, R.; Favati, F.; Pasini, G. Pasta fortification with olive pomace: Effects on the technological characteristics and nutritional properties. *LWT Food Sci. Technol.* **2019**, 114. [CrossRef]
6. Rodríguez Montealegre, R.; Romero Peces, R.; Chacón Vozmediano, J.L.; Martínez Gascueña, J.; García Romero, E. Phenolic compounds in skins and seeds of ten grape Vitis vinifera varieties grown in a warm climate. *J. Food Compos. Anal.* **2006**, 19, 687–693. [CrossRef]
7. Cao, H.; Ou, J.; Chen, L.; Zhang, Y.; Szkudelski, T.; Delmas, D.; Daglia, M.; Xiao, J. Dietary polyphenols and type 2 diabetes: Human Study and Clinical Trial. *Crit. Rev. Food Sci. Nutr.* **2019**, 59, 3371–3379. [CrossRef]
8. Rothwell, J.A.; Knaze, V.; Zamora-Ros, R. Polyphenols: Dietary assessment and role in the prevention of cancers. *Curr. Opin. Clin. Nutr. Metab. Care* **2017**, 20, 512–521. [CrossRef]
9. Zhang, H.; Wang, H.; Cao, X.; Wang, J. Preparation and modification of high dietary fiber flour: A review. *Food Res. Int.* **2018**, 113, 24–35. [CrossRef]
10. Llobera, A.; Cañellas, J. Dietary fibre content and antioxidant activity of Manto Negro red grape (*Vitis vinifera*): Pomace and stem. *Food Chem.* **2007**, 101, 659–666. [CrossRef]
11. Deng, Q.; Penner, M.H.; Zhao, Y. Chemical composition of dietary fiber and polyphenols of five different varieties of wine grape pomace skins. *Food Res. Int.* **2011**, 44, 2712–2720. [CrossRef]
12. Tseng, A.; Zhao, Y. Wine grape pomace as antioxidant dietary fibre for enhancing nutritional value and improving storability of yogurt and salad dressing. *Food Chem.* **2013**, 138, 356–365. [CrossRef] [PubMed]
13. Jane, M.; McKay, J.; Pal, S. Effects of daily consumption of psyllium, oat bran and polyGlycopleX on obesity-related disease risk factors: A critical review. *Nutrition* **2019**, 57, 84–91. [CrossRef] [PubMed]
14. Gawlik-Dziki, U.; Dziki, D.; Świeca, M.; Sęczyk, Ł.; Rózyło, R.; Szymanowska, U. Bread enriched with Chenopodium quinoa leaves powder—The procedures for assessing the fortification efficiency. *LWT Food Sci. Technol.* **2015**, 62, 1226–1234. [CrossRef]

15. Bolarinwa, I.F.; Aruna, T.E.; Raji, A.O. Nutritive value and acceptability of bread fortified with moringa seed powder. *J. Saudi Soc. Agric. Sci.* **2019**, *18*, 195–200. [CrossRef]

16. Fagundes, G.A.; Rocha, M.; Salas-Mellado, M.M. Improvement of protein content and effect on technological properties of wheat. *Food Res.* **2018**, *2*, 221–227. [CrossRef]

17. Reshmi, S.K.; Sudha, M.L.; Shashirekha, M.N. Starch digestibility and predicted glycemic index in the bread fortified with pomelo (*Citrus maxima*) fruit segments. *Food Chem.* **2017**, *237*, 957–965. [CrossRef]

18. Colantuono, A.; Ferracane, R.; Vitaglione, P. Potential bioaccessibility and functionality of polyphenols and cynaropicrin from breads enriched with artichoke stem. *Food Chem.* **2018**, *245*, 838–844. [CrossRef]

19. Mainente, F.; Menin, A.; Alberton, A.; Zoccatelli, G.; Rizzi, C. Evaluation of the sensory and physical properties of meat and fish derivatives containing grape pomace powders. *Int. J. Food Sci. Technol.* **2019**, *54*, 952–958. [CrossRef]

20. Khoozani, A.A.; Kebede, B.; Birch, J.; El-Din Ahmed Bekhit, A. The effect of bread fortification with whole green banana flour on its physicochemical, nutritional and in vitro digestibility. *Foods* **2020**, *9*, 152. [CrossRef]

21. El-Gammal, R.; Ghoneim, G.; ElShehawy, S. Effect of Moringa Leaves Powder (Moringa oleifera) on Some Chemical and Physical Properties of Pan Bread. *J. Food Dairy Sci.* **2016**, *7*, 307–314. [CrossRef]

22. Seguchi, M.; Morimoto, N.; Abe, M.; Yoshino, Y. Effect of maitake (*Grifola frondosa*) mushroom powder on bread properties. *J. Food Sci.* **2001**, *66*, 261–264. [CrossRef]

23. Miranda-Ramos, K.C.; Sanz-Ponce, N.; Haros, C.M. Evaluation of technological and nutritional quality of bread enriched with amaranth flour. *LWT Food Sci. Technol.* **2019**, *114*, 108418. [CrossRef]

24. Hayta, M.; Özuğur, G.; Etgü, H.; Şeker, İ.T. Effect of grape (*Vitis vinifera* L.) pomace on the quality, total phenolic content and anti-radical activity of bread. *J. Food Process. Preserv.* **2014**, *38*, 980–986. [CrossRef]

25. Ahmed, J.; Almusallam, A.S.; Al-Salman, F.; AbdulRahman, M.H.; Al-Salem, E. Rheological properties of water insoluble date fiber incorporated wheat flour dough. *LWT Food Sci. Technol.* **2013**, *51*, 409–416. [CrossRef]

26. Cisneros-Yupanqui, M.; Zagotto, A.; Alberton, A.; Lante, A.; Zagotto, G.; Ribaudo, G.; Rizzi, C. Monitoring the antioxidant activity of an eco-friendly processed grape pomace along the storage. *Nat. Prod. Res.* **2020**, 1–4. [CrossRef] [PubMed]

27. AACC. *Approved Method of the AACC*, 10th ed.; American Association of Cereal Chemists, Ed.; American Association of Cereal Chemists: St. Paul, MN, USA, 2000.

28. Texture Technologies Corp. Texture Analysis Procedures. In *AIB Standard Procedures*; Texture Technologies Corp.: Kansas City, MO, USA, 2011.

29. AOAC. *Official Methods of Analysis*, 18th ed.; The Association of Official Analytical Chemist, Ed.; The Association of Official Analytical Chemist: Gaithersburg, MD, USA, 2007.

30. Del Pino-García, R.; González-SanJosé, M.L.; Rivero-Pérez, M.D.; García-Lomillo, J.; Muñiz, P. The effects of heat treatment on the phenolic composition and antioxidant capacity of red wine pomace seasonings. *Food Chem.* **2017**, *221*, 1723–1732. [CrossRef]

31. Simonato, B.; Tolve, R.; Rainero, G.; Rizzi, C.; Sega, D.; Rocchetti, G.; Lucini, L.; Giuberti, G. Technological, nutritional, and sensory properties of durum wheat fresh pasta fortified with *Moringa oleifera* L. leaf powder. *J. Sci. Food Agric.* **2020**. [CrossRef]

32. Benzie, I.F.F.; Strain, J.J. The Ferric Reducing Ability of Plasma (FRAP) as a Mesure of "Antioxidant Power": The FRAP Assay. *Anal. Biochem.* **1996**, *239*, 70–76. [CrossRef]

33. Ballus, C.A.; Meinhart, A.D.; De Souza Campos, F.A.; Godoy, H.T. Total phenolics of virgin olive oils highly correlate with the hydrogen atom transfer mechanism of antioxidant capacity. *JAOCS J. Am. Oil Chem. Soc.* **2015**, *92*, 843–851. [CrossRef]

34. Vilanova, M.; Masa, A.; Tardaguila, J. Evaluation of the aroma descriptors variability in Spanish grape cultivars by a quantitative descriptive analysis. *Euphytica* **2009**, *165*, 383–389. [CrossRef]

35. Mironeasa, S.; Codină, G.G.; Popa, C. Effect of the addition of Psyllium fiber on wheat flour dough rheological properties. *Recent Res. Med. Biol. Biosci.* **2013**, *XII*, 49–53.

36. Anil, M. Using of hazelnut testa as a source of dietary fiber in breadmaking. *J. Food Eng.* **2007**, *80*, 61–67. [CrossRef]

37. Šporin, M.; Avbelj, M.; Kovač, B.; Možina, S.S. Quality characteristics of wheat flour dough and bread containing grape pomace flour. *Food Sci. Technol. Int.* **2018**, *24*, 251–263. [CrossRef] [PubMed]

38. Girard, A.L.; Awika, J.M. Effects of edible plant polyphenols on gluten protein functionality and potential applications of polyphenol–gluten interactions. *Compr. Rev. Food Sci. Food Saf.* **2020**, *19*, 2164–2199. [CrossRef]

39. Davoudi, Z.; Shahedi, M.; Kadivar, M. Effects of pumpkin powder addition on the rheological, sensory, and quality attributes of Taftoon bread. *Cereal Chem.* **2020**, *97*, 904–911. [CrossRef]

40. Gomez, M.; Oliete, B.; Caballero, P.A.; Ronda, F.; Blanco, C.A. Effect of nut paste enrichment on wheat dough rheology and bread volume. *Food Sci. Technol. Int.* **2008**, *14*, 57–65. [CrossRef]

41. Belghith Fendri, L.; Chaari, F.; Maaloul, M.; Kallel, F.; Abdelkafi, L.; Ellouz Chaabouni, S.; Ghribi-Aydi, D. Wheat bread enrichment by pea and broad bean pods fibers: Effect on dough rheology and bread quality. *LWT Food Sci. Technol.* **2016**, *73*, 584–591. [CrossRef]

42. Pasqualone, A.; Caponio, F.; Simeone, R. Quality evaluation of re-milled durum wheat semolinas used for bread-making in Southern Italy. *Eur. Food Res. Technol.* **2004**, *219*, 630–634. [CrossRef]

43. Pasqualone, A.; Laddomada, B.; Spina, A.; Todaro, A.; Guzmàn, C.; Summo, C.; Mita, G.; Giannone, V. Almond by-products: Extraction and characterization of phenolic compounds and evaluation of their potential use in composite dough with wheat flour. *LWT Food Sci. Technol.* **2018**, *89*, 299–306. [CrossRef]

44. Mehfooz, T.; Mohsin Ali, T.; Arif, S.; Hasnain, A. Effect of barley husk addition on rheological, textural, thermal and sensory characteristics of traditional flat bread (chapatti). *J. Cereal Sci.* **2018**, *79*, 376–382. [CrossRef]

45. Mironeasa, S.; Zaharia, D.; Codină, G.G.; Ropciuc, S.; Iuga, M. Effects of Grape Peels Addition on Mixing, Pasting and Fermentation Characteristics of Dough from 480 Wheat Flour Type. *Bull. UASVM Food Sci. Technol.* **2018**, *75*, 27. [CrossRef]

46. Hammed, A.M.; Ozsisli, B.; Simsek, S. Utilization of Microvisco-Amylograph to Study Flour, Dough, and Bread Qualities of Hydrocolloid/Flour Blends. *Int. J. Food Prop.* **2016**, *19*, 591–604. [CrossRef]

47. Parafati, L.; Restuccia, C.; Palmeri, R.; Fallico, B.; Arena, E. Characterization of prickly pear peel flour as a bioactive and functional ingredient in bread preparation. *Foods* **2020**, *9*, 1189. [CrossRef] [PubMed]

48. Ni, Q.; Ranawana, V.; Hayes, H.E.; Hayward, N.J.; Stead, D.; Raikos, V. Addition of broad bean hull to wheat flour for the development of high-fiber bread: Effects on physical and nutritional properties. *Foods* **2020**, *9*, 1192. [CrossRef]

49. Gonzalez, M.; Reyes, I.; Carrera-Tarela, Y.; Vernon-Carter, E.J.; Alvarez-Ramirez, J. Charcoal bread: Physicochemical and textural properties, in vitro digestibility, and dough rheology. *Int. J. Gastron. Food Sci.* **2020**, *21*, 100227. [CrossRef]

50. Nakov, G.; Brandolini, A.; Hidalgo, A.; Ivanova, N.; Stamatovska, V.; Dimov, I. Effect of grape pomace powder addition on chemical, nutritional and technological properties of cakes. *LWT* **2020**, *134*, 109950. [CrossRef]

51. Arendt, E.K.; Ryan, L.A.M.; Dal Bello, F. Impact of sourdough on the texture of bread. *Food Microbiol.* **2007**, *24*, 165–174. [CrossRef]

52. Zhang, Y.; Hong, T.; Yu, W.; Yang, N.; Jin, Z.; Xu, X. Structural, thermal and rheological properties of gluten dough: Comparative changes by dextran, weak acidification and their combination. *Food Chem.* **2020**, *330*, 127154. [CrossRef]

53. Narendranath, N.V.; Power, R. Relationship between pH and medium dissolved solids in terms of growth and metabolism of lactobacilli and Saccharomyces cerevisiae during ethanol production. *Appl. Environ. Microbiol.* **2005**, *71*, 2239–2243. [CrossRef]

54. Fu, J.T.; Chang, Y.H.; Shiau, S.Y. Rheological, antioxidative and sensory properties of dough and Mantou (steamed bread) enriched with lemon fiber. *LWT Food Sci. Technol.* **2015**, *61*, 56–62. [CrossRef]

55. Sęczyk, Ł.; Świeca, M.; Dziki, D.; Anders, A.; Gawlik-Dziki, U. Antioxidant, nutritional and functional characteristics of wheat bread enriched with ground flaxseed hulls. *Food Chem.* **2017**, *214*, 32–38. [CrossRef] [PubMed]

56. Sivam, A.S.; Sun-Waterhouse, D.; Quek, S.; Perera, C.O. Properties of bread dough with added fiber polysaccharides and phenolic antioxidants: A review. *J. Food Sci.* **2010**, *75*, R163–R174. [CrossRef] [PubMed]

57. Sui, X.; Zhang, Y.; Zhou, W. Bread fortified with anthocyanin-rich extract from black rice as nutraceutical sources: Its quality attributes and in vitro digestibility. *Food Chem.* **2016**, *196*, 910–916. [CrossRef] [PubMed]

58. Begum, Y.A.; Chakraborty, S.; Deka, S.C. Bread fortified with dietary fibre extracted from culinary banana bract: Its quality attributes and in vitro starch digestibility. *Int. J. Food Sci. Technol.* **2020**, *55*, 2359–2369. [CrossRef]

59. Foschia, M.; Peressini, D.; Sensidoni, A.; Brennan, C.S. The effects of dietary fibre addition on the quality of common cereal products. *J. Cereal Sci.* **2013**, *58*, 216–227. [CrossRef]

60. Rizzello, C.G.; Lorusso, A.; Montemurro, M.; Gobbetti, M. Use of sourdough made with quinoa (*Chenopodium quinoa*) flour and autochthonous selected lactic acid bacteria for enhancing the nutritional, textural and sensory features of white bread. *Food Microbiol.* **2016**, *56*, 1–13. [CrossRef]

61. Simonato, B.; Pasini, G.; Giannattasio, M.; Peruffo, A.D.B.; De Lazzari, F.; Curioni, A. Food allergy to wheat products: The effect of bread baking and in vitro digestion on wheat allergenic proteins. A study with bread dough, crumb, and crust. *J. Agric. Food Chem.* **2001**, *49*, 5668–5673. [CrossRef]

62. Hoye, C.; Ross, C.F. Total Phenolic Content, Consumer Acceptance, and Instrumental Analysis of Bread Made with Grape Seed Flour. *J. Food Sci.* **2011**, *76*. [CrossRef]

63. Walker, R.; Tseng, A.; Cavender, G.; Ross, A.; Zhao, Y. Physicochemical, Nutritional, and Sensory Qualities of Wine Grape Pomace Fortified Baked Goods. *J. Food Sci.* **2014**, *79*, S1811–S1822. [CrossRef]

Article

High-Quality Gluten-Free Sponge Cakes without Sucrose: Inulin-Type Fructans as Sugar Alternatives

Urszula Krupa-Kozak [1,*], Natalia Drabińska [1,2], Cristina M. Rosell [2], Beata Piłat [3], Małgorzata Starowicz [1], Tomasz Jeliński [4] and Beata Szmatowicz [5]

[1] Department of Chemistry and Biodynamics of Food, Institute of Animal Reproduction and Food Research of Polish Academy of Sciences, 10-748 Olsztyn, Poland; n.drabinska@pan.olsztyn.pl (N.D.); m.starowicz@pan.olsztyn.pl (M.S.)

[2] Food Science Department, Institute of Agrochemistry and Food Technology (IATA-CSIC), Paterna, 46980 Valencia, Spain; crosell@iata.csic.es

[3] Chair of Food Plant Chemistry and Processing, University of Warmia and Mazury in Olsztyn, 10-748 Olsztyn, Poland; beata.pilat@uwm.edu.pl

[4] Department of Chemical and Physical Properties of Food, Institute of Animal Reproduction and Food Research of Polish Academy of Sciences, 10-748 Olsztyn, Poland; t.jelinski@pan.olsztyn.pl

[5] Sensory Laboratory, Institute of Animal Reproduction and Food Research of Polish Academy of Sciences, 10-748 Olsztyn, Poland; b.szmatowicz@pan.olsztyn.pl

* Correspondence: u.krupa-kozak@pan.olsztyn.pl; Tel.: +48-(89)-523-4618

Received: 16 October 2020; Accepted: 24 November 2020; Published: 25 November 2020

Abstract: Due to its structural and organoleptic functions, sucrose is one of the primary ingredients of many baked confectionery products. In turn, the growing awareness of the association between sugar overconsumption and the development of chronic diseases has prompted the urgent need to reduce the amount of refined sugar in foods. This study aimed to evaluate the effect of complete sucrose replacement with inulin-type fructans (ITFs), namely fructooligosaccharide (FOS), inulin (INU) or oligofructose-enriched inulin (SYN), with different degrees of polymerization on the technological parameters and sensory quality of gluten-free sponge cakes (GFSs). The use of ITFs as the sole sweetening ingredient resulted in the similar appearance of the experimental GFSs to that of the control sample. In addition, all GFSs containing ITFs had similar height, while their baking weight loss was significantly ($p < 0.05$) lower compared to the control products. The total sugar exchange for long-chain INU increased the crumb hardness, while the crumb of the GFS with FOS was as soft as of the control products. The sensory analysis showed that the GFS containing FOS obtained the highest scores for the overall quality assessment, similar to the sugar-containing control sponge cake. The results obtained prove that sucrose is not necessary to produce GFSs with appropriate technological parameters and a high sensory quality. Thus, it can be concluded that sucrose can be successfully replaced with ITF, especially with FOS, in this type of baked confectionery product.

Keywords: sucrose replacement; cake; dietary fibre; clean label; texture profile; sensory quality; obesity; celiac disease

---

1. Introduction

Sugar is one of the primary ingredients of many baked confectionery products, including sponge cakes [1]. Due to many functions critical to obtain the desirable structure and organoleptic properties, sucrose is the most commonly used sugar. It imparts a clean, sweet taste appreciated by consumers, and increases the temperature of starch gelatinization and egg protein denaturation, allowing gas bubbles to expand before gel formation [2]. In addition, sucrose provides foam stability and extends the cake shelf-life [1]. When exposed to the high temperature, sucrose degrades to fructose and glucose, which are the reducing sugars participating in the Maillard browning reactions. In turn, ample studies have provided the evidence for the adverse effects of free sugars overconsumption on health, in particular on the risk of development of non-communicable diseases [3], adverse changes in serum lipids and blood pressure [4], and even cancer [5]. Recently, excessive sugar consumption has attracted particular attention from researchers and epidemiologists, who perceive this phenomenon as a major contributor to the rise in obesity prevalence [6–8].

The global prevalence of obesity has increased to pandemic proportions [9]. This worrisome trend is progressively reported in subjects suffering from celiac disease (CD), which changes the clinical picture of this disorder. Tucker et al. [10] found that at the time of diagnosis 44% of adult CD were overweight, 13% were obese, whereas only 3% of them were underweight. Besides, endocrine autoimmunity, particularly in type 1 diabetes, is prevalent among CD patients, approximating 5–7% [11]. The strict gluten-free diet (GFD) is the only available and effective CD therapy. However, health-care professionals have problems with the optimal approach to treating CD in type 1 diabetes [12]. A GFD alleviates the clinical symptoms and improves the health and nutritional status of CD patients over time. On the other side, it is less clear if the strict adherence to the GFD is similarly important to asymptomatic CD patients with concomitant type 1 diabetes [13], as there is scarce unbiased evidence regarding the influence of a GFD in patients with both autoimmune diseases. Compared with a conventional diet, a GFD is characterised with a lower content of proteins, essential vitamins (B12, D, and folate), and minerals (iron, calcium, zinc) [14,15]. In turn, numerous studies have provided evidence for nutrient imbalance resulting from the excessive consumption of hypercaloric and hyperlipidemic packaged gluten-free foods [16]. Many gluten-free products are abundant in simple sugars and saturated fats [17], which are added to improve their palatability and texture. In contrast, the GFD has been reported to provide a lower than recommended intake of dietary fibre [18], having a negative health consequence.

The growing awareness of the association between excessive sucrose intake and development of chronic diseases has prompted an urgent need to reformulate foods to reduce the amount of refined sugar. On the other hand, the numerous advantages and favourable functional properties of sucrose make its total replacement a challenge. The 100% sugar removal caused readily detectable losses in the appearance, texture, and mouthfeel of baked confectionery products [19]. The synthetic low-calorie sweeteners have attracted consumers' attention and became readily available [20]. However, apart from the high-intensity sweetness, they usually do not provide other functionalities of sucrose necessary to make high-quality cakes. In addition, their breakdown products have controversial health and metabolic effects [21,22].

Recently, there has been an increasing consumer interest in foods of superior quality made from natural ingredients providing functional characteristics while having a reduced sugar content and a lower energy value. As sucrose—being a principal cake ingredient—cannot be easily substituted only by intense sweeteners, several studies have explored the application of natural bulking agents, including dietary fibres, in combination with sweeteners in different cake formulations. Psimouli et al. [2] investigated whether oligofructose, polydextrose, and polyols can replace sugar in cake formulations and showed that oligofructose, lactitol, or maltitol exhibited behaviour similar to sucrose in terms of their influence on batter rheology. To evaluate whether steviol glycosides could partially replace sucrose in bakery products, Zahn et al. [23] produced muffins where 30% sucrose was replaced with rebaudioside A together with several fibres and indicated that a mixture of inulin or polydextrose with steviol glycosides resulted in products with characteristics similar to those of the control muffins. Similarly, Gao et al. [24] used inulin (Frutafit IQ, DP_{av} 5–7, Sensus, Roosendaal, The Netherlands) and stevianna as sucrose substitutes in muffin formulation. They pointed out that the replacement of 50% sugar resulted in muffins having texture, firmness, and springiness similar to these of the control products, while the increased additions of stevianna or inulin had a negative effect on muffin firmness.

In order to fulfil the demand for healthier sponge cakes, studies on sucrose replacement/reduction have been carried out and the use of dietary fibres in combination with sweeteners was proposed.

Ronda et al. [19] analysed the effects of total sugar substitution with polysaccharides, oligofructose, and polydextrose, in combunation with polyols on the quality of sponge cake. They indicated that the fresh sponge cakes with polyols and oligosaccharides generally had significantly softer crumb than the control ones; however, the use of oligofructose (Raftilose P-95, Orafti Active Food Ingredients, Oreye, Belgium) caused crust darkening and an increase in crumb firmness during storage. A recent study by Garvey et al. [25] has explored the impact of partial (30%; *w/w*) sucrose replacement with natural sweetening ingredients, including oligofructose, in sponge cakes and showed that in comparison to control sample, the formula with oligofructose was not significantly different in terms of the liking of colour, odour, flavour, texture, and overall liking.

To successfully exchange sucrose in the already challenging gluten-free system [26] and obtain a desirable structure and organoleptic properties, the sugar replacing ingredient must exhibit the ability to mimic the functionality of sucrose. Due to their sweetness and beneficial technological and health-related properties, inulin-type fructans (ITFs) could be valuable ingredients of gluten-free products [27,28]. Inulin-type fructans are plant natural storage carbohydrates that occur in many edible fruits and vegetables, and in particularly large amounts in the tubers of Jerusalem artichoke (*Helianthus tuberosus*) and chicory (*Cichorium intybus*). They can be divided into long-chain inulin and short-chain fructooligosaccharides (FOS). The length of the chain determines the physicochemical properties of ITFs [29]. Short-chain FOS (DP < 10) are more soluble and sweeter; therefore, they could be used to improve the mouthfeel of low-caloric products [30], while inulin (DP > 10), due to its lower solubility, higher viscosity and thermostability, could be used as a filler and fat-replacer [31]. Investigations showing the use of ITFs in the baked gluten-free products are scarce and mainly focused on gluten-free bread. To the best of our knowledge, this work represents the first study on the ITFs application as natural sugar alternatives in the gluten-free sponge cake formulation. The study aimed to evaluate the effect of total sucrose replacement with the commercial ITFs of different sweetness and degrees of polymerization, namely fructooligosaccharide, inulin, or oligofructose-enriched inulin, on the mixing and pasting batter behaviour and the quality of gluten-free mini-sponge cakes (GFSs) assessed based on selected technological parameters and sensory descriptors.

2. Materials and Methods

2.1. Ingredients of Gluten-Free Mini-Sponge Cakes

The ingredients used to make gluten-free mini-sponge cakes (GFSs) are shown in Table 1. Potato starch (PPZ "Trzemeszno" Sp. Z o.o., Trzemeszno, Poland), corn starch (HORTIMEX, Konin, Poland), and fresh eggs from the local supermarket were the main components. The remaining ingredients were rapeseed oil "Kujawski" (ZT "Kruszwica" S.A., Kruszwica, Poland), gluten-free baking powder (BEZGLUTEN, Koniusza, Poland), sugar, and salt.

Table 1. Composition of experimental gluten-free mini-sponge cakes.

Ingredient [%]	Control	FOS	INU	SYN
Potato starch	30.6	30.6	30.6	30.6
Corn starch	7.8	7.8	7.8	7.8
Egg	43.0	43.0	43.0	43.0
Sugar	14.0	-	-	-
FOS	-	14.0	-	-
INU	-	-	14.0	-
SYN	-	-	-	14.0
Sunflower oil	3.7	3.7	3.7	3.7
Salt	0.2	0.2	0.2	0.2
Gluten-free baking powder	0.7	0.7	0.7	0.7

FOS—fructooligosaccharides; INU—inulin; SYN—oligofructose-enriched inulin.

To produce experimental gluten-free mini-sponge cakes, sugar in the control GFS composition was totally replaced with one of the three commercial inulin-type fructans (ITFs), namely fructooligosaccharide (FOS) with DP_{av} 2–8 and 30% sweetness compared to sucrose (Orafti® P95, Beneo, Tienen, Belgium), inulin (INU) with DP_{av} 8–13 and 10% sweetness compared to sucrose (Frutafit HD, Sensus, Roosendaal, The Netherlands), or oligofructose-enriched inulin (SYN), which is a mixture of oligofructose (DP_{av}: 3–9) and inulin ($DP_{av} \geq 10$) at a specific ratio of 1:1, and ~25% sweetness compared to sucrose (Orafti® Synergy 1, Beneo, Tienen, Belgium), according to product specification.

2.2. Preparation of Experimental Gluten-Free Mini-Sponge Cakes

Gluten-free mini-sponge cakes were prepared following the previously developed method [32]. Briefly, egg whites and salt were whisked (2 min) to form a foam in the stainless bowl in the mixer (KitchenAid Professional K45SS, KitchenAid Europa, Inc., Brussels, Belgium). Then, egg yolks and sugar (in the control GFS) or ITFs (in GFSs with ITFs) were added under continuous vigorous mixing (3 min). Subsequently, starches, baking powder, and oil were added and mixed (3 min/minimal speed) to obtain a smooth homogenous batter. The 30 g portions of batter were dosed into paper moulds (diameter: 50 mm, high: 35 mm), that were put on a baking tray (arranged in three rows, each of four GFSs) and baked at 180 °C for 25 min in an electric oven (AB model DC-21, SVEBA DAHLEN, Fristad, Sweden). Baked GFSs were cooled for 1 h at the room temperature, then packed in a clip-on polyethylene bags, and stored at room temperature pending further analysis. The baking weight loss and height determinations, and instrumental colour analysis of GFSs were performed after cooling (1 h at room temperature), while texture profile and sensory analysis were performed on GFSs stored for 1 day under the described conditions.

2.3. Pasting Behaviour of Batters for Gluten-Free Mini-Sponge Cakes over Heating-Cooling Cycles Determined with the Rapid Visco Analyser (RVA)

The viscosity of batters for gluten-free mini-sponge cakes over heating-cooling cycles was evaluated using a Rapid Visco Analyser (RVA-4800; Perten Instruments, Madrid, Spain). The GFS batters were prepared according to the formula presented in Table 1, with the exception that fresh eggs were replaced with whole egg powder (EPSA Additivos Alimentarios, Valencia, Spain) to avoid the variability that the fresh eggs could introduce in a collaborative project. The 8 g batter samples were dispersed in distilled water (12 mL). The obtained suspensions were stirred for 1 min at 600 rpm at 30 °C. After that time, the temperature rose to 95 °C at a rate of 12 °C min^{-1}. The sample was maintained for 30 s at 95 °C, cooled to 50 °C at a rate of 12 °C min^{-1}, and finally maintained for 2 min at 50 °C. The onset temperature (°C), peak temperature (°C), peak viscosity (PV, cP), hot paste viscosity (HPV; cP), breakdown (PV-HPV; cP), cold paste (final) viscosity (CPV; cP), and setback (CPV-HPV; cP) were recorded. The experiments were conducted in triplicate.

2.4. Characteristics of Gluten-Free Mini-Sponge Cakes

2.4.1. Physical Parameters

The weight of GFSs was measured using a digital balance. The height of GFSs was measured at the highest point of the product using a digital calliper. Baking weight loss (WL) was calculated as the ratio between the weight of batter and the weight of the baked and cooled GFS (Equation (1)):

$$WL\ (\%) = \frac{(a - c) \times 100}{a}$$

(1)

where:
   a—the weight of batter in the mould before baking (g),
   c—the weight of baked and cooled GFS (g).

2.4.2. Texture Profile Analysis

Textural properties of experimental GFSs were assessed 24 h after baking. The samples were removed from clip-on bags just before testing. Like in the previous study [32], the GFSs were cut horizontally at the height of the mould to form a flat surface. The texture profile analysis (TPA) was performed on the 2 cm-high lower part whereas the upper part was discarded. Hardness, springiness, gumminess, chewiness, cohesiveness, and resilience were determined using a TA.HD Plus Texture Analyser (Stable Micro Systems Ltd., Godalming, UK) equipped with a 5-kg load cell. The sample was placed centrally under an AACC 36-mm cylinder probe with radius (P/36R). The GFS sample was compressed at a constant rate of 1.0 mm s^{-1} at a distance of 5 mm. The probe holds at this distance for 30 s and then withdraws from the sample and returns to its starting position. Each type of GFSs was tested in six replications.

2.4.3. Instrumental Colour Analysis

The instrumental measurements of the crust and crumb colour of GFSs were made using a HunterLab ColorFlex (Hunter Associates Laboratory, Inc., Reston, VA, USA). The measurements were performed through a 3-cm-diameter diaphragm containing an optical glass. The results were expressed in accordance with the CIELab system. The parameters determined were L^* ($L^* = 0$ (black) and $L^* = 100$ (white)), a^* ($+a^*$ = redness and $-a^*$ = greenness), and b^* ($+b^*$ = yellowness and $-b^*$ = blueness). Values were the mean of at least six replicates.

The whiteness index (WI) of the crumb [33] was calculated according to Equation (2):

$$\text{WI} = 100 - \sqrt{(100 - L^*)^2 + a^{*2} + b^{*2}} \tag{2}$$

The browning index (BI) of the crust [34] was calculated according to Equations (3) and (4):

$$\text{BI} = \frac{100 \times (x - 0.31)}{0.17} \tag{3}$$

where:

$$x = \frac{a^* + 1.75L^*}{5.645L^* + a^* - 3.012b^*} \tag{4}$$

The ΔE_{Lab} difference between two colours [35] was calculated according to Equation (5):

$$\Delta E_{Lab} = \sqrt{(\Delta L^*)^2 + (\Delta a^*)^2 + (\Delta b^*)^2} \tag{5}$$

2.4.4. Evaluation of Early, Advanced, and Final Stage of the Maillard Reaction

The content of available lysine, as an indicator of the early stage of the Maillard reaction, was determined according to the method described by Michalska et al. [36]. Exactly 50 µL of a sample, 100 µL of o-phthaldialdehyde reagent, and 100 µL of water were added to wells and incubated for 3 min (96-well microplate; Porvair Sciences, Norfolk, UK). Then, fluorescence was measured at $\lambda_{extinction} = 340$ nm and $\lambda_{emmision} = 455$ nm using a microplate reader (Infinite® M1000 PRO, Tecan, Switzerland). The quantitative analysis was performed according to the external standard method, employing a calibration curve of N_α-acetyl-L-lysine ranging from 10 to 250 µM. The content of free intermediate compounds (FIC) was determined after sample extraction with 6% sodium dodecyl sulfate and then their fluorescence was recorded in a microplate reader (Infinite® M1000 PRO, Tecan, Switzerland) setting at $\lambda_{extinction} = 347$ nm and $\lambda_{emmision} = 415$ nm. Tryptophan fluorescence (TRP) was measured at $\lambda_{extinction} = 290$ nm and $\lambda_{emmision} = 340$ nm. Results are expressed as fluorescence intensity (FI) per mg of sample dry matter. The FIC and FAST (fluorescence of advanced MRPs and tryptophan) index were calculated as recently reported by Zieliński et al. [37]. The FAST index data were expressed as a percentage (%). The formation of brown pigments (melanoidins) was estimated as reported in detail by Zieliński et al. [37]. Results were expressed as arbitrary absorbance units. All measurements were performed in triplicate.

2.4.5. Sensory Evaluation

A six-member expert panel (five women and one men) previously selected and trained according to ISO guidelines [38] evaluated the sensory characteristics of experimental GFSs 24 h after baking. The assessors were not CD patients but were familiar with gluten-free products and have been aware of tasting starch-based gluten-free sponge cake. A quantitative descriptive analysis (QDA) [39] was applied to assess the sensory characteristics of the experimental GFSs. Before the analysis, vocabularies of the sensory attributes were developed by the panel in a round-table session, using a standardised procedure [40]. Thirteen attributes were evaluated (Table 2). The assessors evaluated the intensity perceived for each sensory attribute on unstructured graphical scales. The scales were 10 cm long and verbally anchored at each end, and the results were converted to numerical values (from 0 to 10 arbitrary units) by a computer. The experimental GFS samples were coded with a three-digit number and presented to the assessors all together in a random order in transparent plastic boxes. The sensory evaluation was carried out in a sensory laboratory room, which fulfils the requirements of the ISO standards [41], under normal lighting conditions at room temperature. To minimise residual effects, bottled mineral water was suggested to drink between each sample evaluation. The results were collected using a computerised system ANALSENS (IAR & FR PAS, Olsztyn, Poland). GFSs were tested in two replications.

Table 2. Sensory attributes, their definitions, and scale edges used in the quantitative descriptive analysis (QDA) of gluten-free mini-sponge cakes.

Attribute		Definition	Scale Edges
Appearance	Creamy colour	colour intensity (colour intensity according to colour pattern RAL 075 90 20—scale value 5)	light–dark
	Pore collocation	a visual impression of the arrangement of crumb pores	irregular–regular
	Pore dimension	a visual impression of the size of crumb pores	small–big
Aroma	Sponge cake	the typical odour of sponge cake	none—very intensive
	Sweet	aroma typical of sweet baked products from wheat flour	none—very intensive
Taste	Sponge cake	as the corresponding odour (measured in the mouth)	none—very intensive
	Sweet	basic taste (3% sucrose dissolved in water)	none—very intensive
	Aftertaste	lingering sensation after swallowing the sample	none—very intensive
Texture (manual)	Elasticity	the extent to which a piece of product returns to its original length when pushed by a finger	small–big
Texture (in mouth)	Crustiness	degree of friability released by the sample	small–big
	Chewiness	the multiplicity of chewing the product to prepare it to swallow	low–high
	Adhesiveness	degree of adhesiveness perceived while chewing the sample 10 times	low–high
Overall quality		overall quality including all attributes and their harmonisation	low–high

2.5. Statistical Analysis

The data reported in all the tables are mean values and standard deviations of triplicate observations unless otherwise stated. The differences between experimental GFSs were analysed by a one-way analysis of variance (ANOVA) with Tukey's multiple comparison test ($p < 0.05$) using GraphPad Prism version 8.0.0 for Windows, GraphPad Software (San Diego, CA, USA).

3. Results and Discussion

3.1. Pasting Behaviour of Batters for Gluten-Free Mini-Sponge Cakes

The analysis of the pasting properties of gluten-free batters is essential in developing high-quality gluten-free products as it provides information about the changes in paste viscosity behaviour with changes in temperature [32,42]. Changes in the pasting of experimental GFS batters recorded as alterations in batter viscosity due to swelling and pasting of starch granules are shown at the RVA plots (Figure 1). In general, the shapes of RVA plots for the control GFS batter (containing sugar) and batters with ITFs (FOS, INU, SYN) did not differ meaningfully at the mixing and initial heating stage (Figure 1) where all pasting curves were characterised with an initial plateau. During this short period, the viscosity of all experimental GFS batters was similarly low. At this stage, the hydration of potato and corn starch granules took place and increased gradually due to the available water that penetrates into the starch's interior. The minimal swelling of starch granules could be observed at these temperature conditions (below 50 °C) [43]. Additionally, at the pasting stage, the behaviour of GFS batters could be affected by egg proteins present in the system [44].

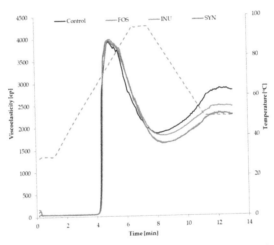

Figure 1. Effect of sucrose replacement with inulin-type fructans (ITFs) on plots of the viscometric profile recorded with the rapid viscoanalyser. FOS—gluten-free mini-sponge cake with FOS; INU—gluten-free mini-sponge cake with inulin; SYN—gluten-free mini-sponge cake with Synergy 1.

Subsequently, as the temperature rose, a sudden increase was recorded in the viscosity of all analysed GFS batters that was observed as a high, sharp, and narrow peak (Figure 1). The starchy ingredients absorbed water available in the batter environment and swelled progressively upon water presence and heat. In all analysed GFS samples, regardless of the presence of sugar (control) or ITFs, excessive expansion of starch granules led to an increase in viscosity up to the maximum apparent viscosity, the so-called peak viscosity (PV). The experimental GFS batters achieved the PV at similar temperature and time (Table 3). Therefore, despite differences in their dextrin chain length, no significant differences were observed in the peak viscosity between the batters. In the next step, when the temperature was constant (95 °C) for 30 s, a substantial reduction in the apparent viscosity of GFS batters was detected (Figure 1) that was determined as the breakdown. Breakdown viscosity is defined as a difference between PV and hot paste viscosity (HPV) and illustrates the ability of the starch to withstand shear stress and heating. Generally, the observed changes were a consequence of the physical breakdown of the starch granules that was accompanied by viscosity decrease. Sugar substitution with ITFs affected this parameter differently; a significant ($p < 0.05$)

reduction was determined in HPV (Table 3), particularly when sucrose was exchanged with FOS and Synergy 1. ITFs with intermediate degrees of polymerization led to the higher breakdown; it is likely that those dextrins affect the amylose leaching that accompanied the gelanization process. In the last stage of the RVA analysis, the final viscosity, determined as cold paste viscosity (CPV), and setback were determined. These parameters reflect the ability of the starch polymers to re-organise when the temperature decreases. The setback viscosity of all experimental GFS batters increased (Table 3), that is commonly related to the crystallisation of the amylose chains, but also to the effect of denatured protein [44]. In comparison with the control batter, the value of setback recorded for the batters containing ITFs was significantly ($p < 0.05$) lower (Table 3), suggesting the lower degree of amylose chains crystallisation. The applied ITFs are soluble fibres, however, they differ in chain length, with inulin having the longest chain. That is why the applied ITFs affected the batter pasting characteristics to a different extent.

Table 3. Effects of sucrose replacement with ITFs on the Rapid Visco Analyser (RVA) parameters of batter of gluten-free mini-sponge cakes.

	Control	FOS	INU	SYN
Onset temperature (°C) [1]	64 ± 1 [ab]	64 ± 1 [a]	63 ± 1 [b]	64 ± 1 [a]
Peak temperature (°C)	77 ± 1	77 ± 1	76 ± 1	77 ± 1
Peak viscosity PV (cP)	3937 ± 16	4014 ± 80	3975 ± 122	3973 ± 54
HPV (cP)	1856 ± 15 [a]	1648 ± 54 [b]	1808 ± 43 [a]	1633 ± 61 [b]
Breakdown (cP)	2080 ± 1 [b]	2366 ± 26 [a]	2167 ± 84 [b]	2339 ± 15 [a]
Final CPV (cP)	2846 ± 50 [a]	2253 ± 68 [c]	2445 ± 67 [b]	2265 ± 57 [c]
Setback (cP)	991 ± 64 [a]	605 ± 14 [c]	637 ± 30 [b]	632 ± 5 [c]

[1] Values were presented as mean ($n = 3$) ± standard deviation. [a–c] Means with different letters in the same row are significantly different ($p < 0.05$), as determined by Tukey's multiple comparisons test. HPV—hot paste viscosity; CPV—cold paste viscosity.

The study confirmed that the presence of different ITFs in a cake batter modified the pasting profile, particularly after heating and cooling. It has been reported that soluble and insoluble fibres affect the performance of gluten-free layer cakes batter [45], showing that fibres increased the batter viscosity, except for inulin, which decreased it. The present study results even show that the behaviour of the inulins was greatly dependent on their degree of polymerization.

3.2. Physical Characteristics and Texture of Gluten-Free Mini-Sponge Cakes

The effect of sugar replacement with ITFs on the physical characteristics of GFSs is shown in Table 4 and Figure 2. The experimental sponge cakes containing ITFs were significantly ($p < 0.05$) heavier but similarly high as the control GFS with sugar (Table 4). In the case of foam-type cakes, including particulalry sponge cakes, the high volume and fine porosity are desirable features [1]. The elimination of sucrose did not affect GFSs' volume. Thus, it appears that ITFs consolidated the structure of experimental GFSs, supported the height, and prevented GFSs' collapse after baking (Figure 2). The height and final volume of foam-type cakes are mainly determined by gas cell incorporation during mixing and steam production during baking [46]; however, the structure of the cake is established during baking together with the rise of the temperature when the cake matrix solidifies as a result of starch gelatinization and protein denaturation.

All experimental GFSs with ITFs were characterised by significantly ($p < 0.05$) lower baking weight loss in comparison with the control cake containing sugar (Table 4). Baking weight loss is one of the major technological losses and therefore efforts are made to minimise it. Generally, a number of physical and chemical modifications proceed during baking, such as evaporation of water, formation of a porous structure, expansion of volume, etc. [47]. The sponge cake baking process can be divided into the heating up period and crust/crumb period [48]. Baking weight loss results mainly from the drying process [49]; however, water vaporization during the initial heating up period may take place as crust does not appear instantaneously. In addition, other ingredients, ITFs in particular, have a great influence on water retention in baked products. In the case of experimental GFSs without sugar, the decreased value of baking weight loss could be due to ITFs' ability to bind water molecules [50], causing the higher water retention in the cake during baking. The results obtained by Rodriguez-Garcia et al. [51], where a highly-dispersible native inulin (Frutafit HD®, average chain length 8–13, Sensus, Roosendaal, The Netherlands) and highly-soluble oligofructose (Frutafit CLR®, average chain length 7–9, Sensus, Roosendaal, The Netherlands) were used indicated that cakes with 50% of native inulin as fat replacer had significantly ($p < 0.05$) lower weight loss than the cakes without it, suggesting that inulins bind water and help to retain moisture during baking.

The effect of sugar replacement with ITFs on textural parameters of the crumb of experimental GFSs is presented in Table 4. The control sponge cake containing sugar had the softest crumb (32.22 N), which at the same time was the most springy and cohesive, and the least gummy and chewy. Sugar replacement with ITFs significantly ($p < 0.05$) affected the TPA profile of the experimental GFSs (Table 4). Inulin increased the hardness of experimental GFSs, while the FOS sample showed crumb softness that was close to the control (37.45 N). Similar observations were made by Gao et al. [24], who revealed that a total replacement of sucrose with inulin (Frutafit IQ, DP_{av} 5–7, Sensus, Roosendaal, The Netherlands) gave muffins with a firmer texture than the control ones. The differences in the action of ITFs applied in the experimental GFSs could be explained by the increase in hardness of inulin gels observed with an increasing degree of polymerisation [52]. In addition, Ziobro et al. [53] demonstrated that the range of changes in textural parameters of the gluten-free bread influenced by ITFs depended on the structure (including DP) and the amount of the applied additives. These authors reported that inulin with a lower DP (HSI with a DP < 10, BENEO-Orafti, Belgium) had a favourable impact on crumb hardness of gluten-free bread, while the loaves with the addition of high performance inulin (HPX) with DP > 23 (BENEO-Orafti, Belgium) were significantly harder. In the case of the remaining texture parameters analysed in the present study, independently of the DP, the complete sugar replacement with ITFs caused rather undesirable changes in the characteristics of GFSs (Table 4). The FOS, INU, and SYN samples were significantly ($p < 0.05$) more gummy and chewy than the control ones, while their springiness and cohesiveness were reduced.

The results of the instrumental colour analysis of crust and crumb of experimental GFSs are shown in Table 4. Compared with the sugar-containing control sponge cake, sugar replacement with ITFs had a significant but different effect on the colour parameters of the crust. Short-chained FOS induced the most pronounced darkening of the crust of the experimental FOS sponge cake ($L^* = 49.53$), followed by SYN. On the other hand, INU in which sugar was replaced with long-chained inulin, the L^* value determined for the crust was significantly ($p < 0.05$) lower (L^* value = 70.69) compared with the control (64.64) and other GFSs with ITFs. Positive values of coordinates a^* (red hue) and b^* (yellow hue) were recorded for the crust of all experimental GFSs (Table 4), regardless of the ITFs applied. The results obtained indicated a reddish shade of crust of all GFSs, with the highest value of coordinate a^* determined in the FOS sample. The value of the b^* coordinate, denoting a yellow shade, was the highest in SYN. The value of the browning index (BI) was inversely related to the crust whiteness; therefore, the highest BI value was recorded for FOS, followed by SYN, whereas the most pronounced reduction in this parameter was recorded for INU (Table 4). Contrary to the crust, the colour of the crumb depends rather on the colour of the ingredients. ITFs used as sugar replacers in the GFS had a similarly white to slightly creamy colour. However, even though no apparent differences were observed in ITFs' colour, they had an impact on the crumb colour of experimental GFSs. The crumb of the FOS sponge cake was whiter than of the control one (Table 4), while regardless the kind of dietary fibre used in the formulation, crumbs of all GFSs were significantly ($p < 0.05$) redder (a^*) and more yellow (b^*), compared with the control products. The crumb of the control sponge cake was characterised by the highest whiteness index (WI) (Table 4). In general, the development of brown colour of food that appears during baking is a very typical phenomenon and is mainly caused by non-enzymatic browning reactions which include, among others, caramelisation and Maillard reactions.

Carbonyl groups of reducing sugars polymerise with α- and ε-amino groups of proteins, peptides, or amino acids to produce brown nitrogenous pigments (melanoidins) by the spontaneous Maillard reaction [54]. The assessment of the crust colour of the experimental GFSs indicated a pronounced darker colour in the sample containing FOS.

Table 4. Effects of sucrose replacement with ITFs on physical characteristics and texture of gluten-free mini-sponge cakes.

	Control	FOS	INU	SYN
Weight (g) [1]	22.45 ± 0.57 [b]	23.13 ± 0.32 [a]	23.31 ± 0.13 [a]	23.23 ± 0.11 [a]
Height (mm)	49.0 ± 0.52	51.3 ± 0.30	51.0 ± 0.36	48.0 ± 0.10
Baking weight loss (%)	25.18 ± 1.92 [a]	22.89 ± 1.05 [b]	22.31 ± 0.44 [b]	22.58 ± 0.37 [b]
Textural parameters				
Hardness (N)	32.22 ± 2.49 [c]	37.45 ± 2.74 [c]	62.92 ± 3.473 [a]	51.54 ± 6.639 [b]
Springiness (%)	0.87 ± 0.03 [a]	0.79 ± 0.02 [b]	0.81 ± 0.02 [b]	0.81 ± 0.03 [b]
Cohesiveness	0.49 ± 0.01 [a]	0.33 ± 0.02 [b]	0.15 ± 0.02 [c]	0.17 ± 0.03 [c]
Gumminess	1.60 ± 0.11 [d]	12.22 ± 0.63 [a]	10.43 ± 0.90 [b]	8.56 ± 0.76 [c]
Chewiness	1.40 ± 0.10 [c]	9.67 ± 0.66 [a]	7.70 ± 0.95 [b]	6.49 ± 0.96 [b]
Resilience	0.15 ± 0.01 [a]	0.08 ± 0.01 [b]	0.05 ± 0.01 [c]	0.06 ± 0.01 [c]
Crust colour				
L^*	64.64 ± 2.82 [b]	49.53 ± 0.12 [d]	70.69 ± 0.95 [a]	59.71 ± 0.47 [c]
a^*	13.26 ± 0.84 [c]	17.64 ± 0.20 [a]	12.09 ± 0.19 [d]	15.72 ± 0.13 [b]
b^*	35.58 ± 0.95 [a]	34.14 ± 0.49 [bc]	33.71 ± 0.54 [c]	36.09 ± 0.19 [a]
BI	92.21 ± 5.94 [c]	133.30 ± 2.56 [a]	75.66 ± 0.70 [d]	107.36 ± 1.46 [b]
ΔE		6.44	15.8	5.53
Crumb colour				
L^*	82.74 ± 0.58 [b]	84.01 ± 0.64 [a]	82.42 ± 0.49 [b]	81.23 ± 0.92 [c]
a^*	2.01 ± 0.11 [d]	3.22 ± 0.14 [b]	2.65 ± 0.14 [c]	4.59 ± 0.19 [a]
b^*	22.75 ± 0.80 [c]	24.95 ± 0.64 [b]	24.15 ± 0.54 [b]	28.84 ± 0.24 [a]
WI	71.37 ± 0.65 [a]	70.18 ± 0.48 [b]	70.01 ± 0.67 [b]	65.28 ± 0.62 [c]
ΔE		1.57	2.82	6.79

[1] Values were presented as mean ± standard deviation. [a–c] Values followed by different letters in the same row are significantly different ($p < 0.05$), as determined by Tukey's multiple comparisons test.

The obtained results of the instrumental colour analysis were consistent with the findings reported by Zahn et al. [31] and suggested that the rate of the Maillard reaction was more intensive in FOS than in other GFSs with ITFs. During the baking process, the hydrolysis of FOS to fructans could occur, thereby increasing the quantity of reducing sugar (especially fructose), promoting the Maillard reaction. The darkening of the crust of experimental GFSs could be perceived as a desirable feature because the gluten-free products generally tend to be paler than their wheat counterparts [26]. In contrast, the light crust of INU may indicate the suppression of the Maillard reaction due to the dilution of the reaction precursor's in the presence of inulin, a water-retaining ingredient, resulting in a higher water content in the environment [55].

Changes in the contents of the early, advanced, and final stage Maillard reaction products affected by sugar replacement with ITFs are presented in Table 5. In this study, the available lysine served as an indicator of the early stage of Maillard reaction, while the fluorescent intermediate products (FIC) formation was considered as the advanced stage of the reaction, and finally, the generation of melanoidins was indicative of the final stage. The FOS sponge cake had available lysine content at the same level as in the control, while that found in INU and SYN was about 20% lower (Table 5). Therefore, the FIC value for control and FOS was similar, whereas the increased amount of FIC was determined in the other GFSs with ITFs. It suggested that INU and SYN promoted the formation of fluorescence compounds, whereas FOS did not. To describe the protein loss in the experimental GFSs, the FAST index was calculated as a ratio between FIC and tryptophan fluorescence presented in%. The lowest value of the FAST index was detected in the control sponge cake (Table 5). Among GFSs with ITFs, a positive effect counteracting proteins loss was noticed in the sample with short-chained FOS, determined as a significantly ($p < 0.05$) lower FAST index percentage. The FAST index values obtained for GFSs were, however, lower than the obtained by Przygodzka et al. [56] for rye-buckwheat cakes enriched with spices. In the GFS containing FOS, an intense form of brown melanoidins was observed (Table 5) that was 45% and 15% higher than in INU and SYN, respectively. The results obtained corresponded well to the results of the instrumental colour analysis (Table 4) and proved that melanoidin formation was positively linked to BI. Nevertheless, except for colour development, many studies demonstrated the health-promoting properties of melanoidins, including their antimicrobial, antioxidant, anti-inflammatory or probiotic effects [57].

Figure 2. Exemplary pictures of the appearance and cross-section of experimental gluten-free mini-sponge cakes with sucrose replaced with ITFs. (**A**) control gluten-free sponge cake; (**B**) gluten-free sponge cake with FOS; (**C**) gluten-free sponge cake with inulin; (**D**) gluten-free sponge cake with Synergy 1.

The acceptance of the sensory quality is essential when a new product is being developed; therefore, an important step in the novel product development is to determine and analyse its quality characteristics, including appearance, aroma, and taste. In the present study, trained experts were asked to assess the experimental GFSs based on their visual appearance (crumb colour and porosity), aroma, taste, and texture, both manually (elasticity) and by the mouth (crustiness chewiness and adhesiveness). The results of QDA are presented in Table 6 and Figure 3. In general, the ITFs used in the formulation did not influence the visual appearance of the crumb of experimental GFSs (Table 6). All GFSs with ITFs looked similar like the creamy-coloured control sponge cake containing sugar. This indicated that the differences in crumb colour detected by the instrumental colour analysis (Table 4) were not perceived by the experts panel in QDA analysis (Table 6). This could be explained by the differences between the methods applied. The instrumental spectrophotometric method makes it possible to define the colour precisely, expressing it numerically in comparison to the standard. The main advantage of this instrumental measurement over the sensory QDA analysis is its higher repeatability resulting from the lower standard deviation due to the lack of variability caused by psychological, physiological, and environmental factors that affect human sensory reactions [58]. The experimental GFSs were similar to the control cakes in terms of porosity features, with both taking into account pore collocation and dimension (Table 6; Figure 2), regardless of the ITFs used in the sponge cake formulation. The number, size, and distribution of air cells incorporated during the mixing stage determine the volume and texture of the baked cakes [1]. A larger number of smaller pores rather than a smaller number of larger ones is a feature of high-quality sponge cakes [32,59]. The control sample was characterised by intensive sponge cake aroma and taste, while in GFSs containing ITFs these features were detected in significantly ($p < 0.05$) lower range, especially in FOS (Table 6). In addition, the experimental GFSs containing ITFs were characterised as having less sweet aroma and taste than the control cakes. Texture evaluation is also an important step in developing a high-quality food product or optimising processing variables.

Table 5. Effects of sucrose replacement with ITFs on the contents of the early, advanced, and final stage Maillard reaction products in gluten-free mini-sponge cakes.

	Control	FOS	INU	SYN
Available lysine (mg/g) [1]	12.50 ± 0.42 [a]	12.51 ± 0.17 [a]	9.88 ± 0.06 [b]	9.31 ± 0.09 [b]
FIC (FI)	28.05 ± 0.60 [bc]	20.16 ± 0.17 [c]	50.14 ± 1.94 [a]	38.56 ± 1.87 [b]
TRP (FI)	18.00 ± 0.09 [d]	16.02 ± 0.03 [c]	20.52 ± 0.78 [a]	19.44 ± 0.04 [b]
FAST index (%)	123 ± 5.56 [c]	175 ± 1.08 [b]	245 ± 11.28 [a]	203 ± 19.94 [b]
Melanoidins (AU)	0.758 ± 0.011 [b]	0.921 ± 0.012 [a]	0.504 ± 0.005 [c]	0.775 ± 0.020 [b]

FIC—free intermediate compounds; TRP—tryptophan fluorescence; FAST index—fluorescence of advanced MRPs and tryptophan. [1] Values were presented as mean ± standard deviation. [a–c] Values followed by different letters in the same row are significantly different ($p < 0.05$), as determined by Tukey's multiple comparisons test.

Table 6. Effects of sucrose replacement with ITFs on the sensory quality assessed with a quantitative descriptive analysis (QDA) in gluten-free mini-sponge cakes.

Attribute		Control	FOS	INU	SYN
Appearance	creamy colour	2.42 ± 0.53	2.59 ± 0.62	2.48 ± 0.54	2.86 ± 0.63
	pore collocation	6.59 ± 1.59	7.16 ± 1.87	7.02 ± 0.83	7.56 ± 1.20
	pore dimension	2.51 ± 1.34	2.20 ± 1.95	2.18 ± 0.47	2.03 ± 0.72
Aroma	sponge cake	8.11 ± 1.07 [a]	2.26 ± 0.85 [d]	5.34 ± 2.04 [b]	3.91 ± 1.32 [c]
	sweet	5.82 ± 1.96 [a]	1.64 ± 0.92 [c]	3.69 ± 2.32 [b]	2.88 ± 1.59 [bc]
Taste	sponge cake	8.12 ± 0.95 [a]	4.04 ± 2.31 [b]	3.34 ± 0.89 [b]	3.05 ± 1.35 [b]
	sweet	6.26 ± 1.48 [a]	1.98 ± 1.20 [b]	1.09 ± 0.76 [b]	0.99 ± 0.80 [b]
	aftertaste	1.39 ± 0.61	1.59 ± 0.96	1.13 ± 0.68	1.31 ± 0.71
Texture (manual)	elasticity	6.18 ± 0.48 [a]	6.69 ± 1.61 [a]	3.04 ± 0.92 [b]	2.92 ± 1.18 [b]
	crustiness	1.58 ± 0.24 [b]	1.77 ± 0.53 [b]	2.72 ± 0.98 [a]	2.76 ± 0.82 [a]
Texture (in mouth)	chewiness	3.54 ± 0.44	3.58 ± 0.54	3.36 ± 0.82	3.43 ± 0.93
	adhesiveness	2.98 ± 0.78	2.67 ± 0.67	2.41 ± 0.76	2.24 ± 0.90

Values were presented as mean ± standard deviation. [a–d] Values followed by different letters in the same row are significantly different ($p < 0.05$), as determined by Tukey's multiple comparisons test.

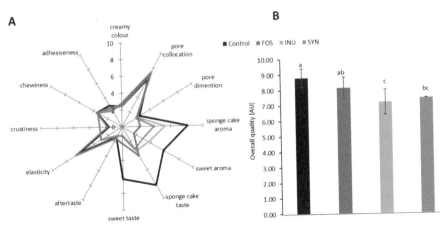

Figure 3. Spider diagram presenting the effect of sucrose replacement with ITFs on QDA parameters (**A**); overall quality (**B**) of gluten-free mini-sponge cakes. FOS—gluten-free mini-sponge cake with FOS; INU—gluten-free mini-sponge cake with inulin; SYN—gluten-free mini-sponge cake with Synergy 1. [a–c] Values followed by different letters above the bars are significantly different ($p < 0.05$), as determined by Tukey's multiple comparisons test.

The ANOVA analysis revealed significant differences ($p < 0.05$) in the QDA texture parameters of the experimental GFSs, in particular in their elasticity (examined manually) and crustiness (assessed in the mouth). In comparison with the control cake, FOS was similarly elastic and had the same crustiness, while INU and SYN were significantly ($p < 0.05$) less elastic and more crusty than the control and FOS (Table 6). The results of sensory analysis corresponded in part with the results of the instrumental texture analysis (Table 4), as both methods indicated a greater similarity of FOS to the control cake than to other GFSs with ITFs. However, the instrumental texture parameters and the sensory descriptors are not defined similarly, while their methodology, including sample size and analysis conditions, are significantly different [60]. When comparing sensory and instrumental analytical methods, it should be noticed that the results of instrumental methods are related to the physical parameters that trigger sensory impressions, while the results of the sensory analysis inform directly about the sensations that these stimuli evoke [61]. Nevertheless, both sensory evaluation techniques and instrumental measurements are equally used to assess texture parameters in food products [62]. In the overall quality assessment, all GFSs containing ITFs were of satisfactory quality, with high scores ranging from 7.21 to 8.13 (Figure 3). However, among experimental GFSs, FOS was favoured and received the highest scores (Figure 2), similar to the control cake containing sugar (8.79). On the other hand, panellists found that the INU and SYN samples were less favoured in the overall acceptance and palatability, compared with the control ($p < 0.05$). These results were in agreement with the results of the instrumental texture and colour analysis (Table 3). The addition of inulin to the experimental sponge cake formulation deteriorated its quality, yielding harder crumb and paler crust (Table 4), and consequently diminishing the sensory quality of GFS. In turn, the use of short-chained FOS improved many technological properties, resulting in the amelioration of the sensorial characteristic of GFS. As was discussed before, the length of the inulin molecule is an important feature affecting the physical and technological properties of the final product. Ziobro et al. [53] reported that the DP of inulin preparations affected the physical characteristics and staling rate of gluten-free bread. Taking into account all gathered results, it could be concluded that a high-quality GFS (i.e., uniform, medium size porosity, proper crumb structure) could be obtained if the appropriate ITFs source was selected based on its functional characteristics, including the DP.

4. Conclusions

Our research has demonstrated the feasibility of complete sucrose replacement with ITFs in the gluten-free sponge cake formulation. The use of ITFs as the sole sweetening ingredient resulted in less sweet GFSs which were, however, characterised by a similar appearance to the control sample, especially in terms of the similar crumb colour and porosity. In addition, all GFSs containing ITFs instead of sugar had similar height, while their baking weight loss was significantly lower than in the control cake. The total sugar replacement with ITFs significantly influenced the texture profile of the experimental GFS, in particular the use of long-chain inulin increased the crumb hardness, while the crumb of the GFS with FOS was as soft as that of the control cake. In comparison with the control sample, the GFS containing FOS had a similar content of early Maillard reaction products, determined as the available lysine, which indicated the favourable counteraction of protein loss in this sample. The QDA analysis showed that among the experimental GFSs with ITFs, the sample containing FOS obtained the highest scores for the overall quality assessment, which was similarly high to that given to the sugar-containing control cake. The results obtained prove that sucrose is not necessary to make a gluten-free sugar-free sponge cake with appropriate technological parameters and high sensory quality. Therefore, it can be successfully replaced with ITF, especially FOS, in this type of baked confectionery product.

Author Contributions: Conceptualization, U.K.-K.; methodology, U.K.-K., C.M.R., B.P. and B.S.; formal analysis, U.K.-K., T.J. and B.S.; investigation, N.D., B.P., M.S., T.J. and B.S.; resources, U.K.-K. and C.M.R.; writing—original draft preparation, U.K.-K.; writing—review and editing, N.D., C.M.R., M.S. and B.S.; visualization, U.K.-K., B.S.; supervision, U.K.-K.; funding acquisition, U.K.-K. All authors have read and agreed to the published version of the manuscript.

Funding: This research was funded by KNOW (Leading National Research Centre) Scientific Consortium: "Healthy Animal—Safe Food"(decision of Ministry of Science and Higher Education No. 05-1/KNOW2/2015 and the APC was partially funded by statutory funds of the Department of Chemistry and Biodynamics of Food of the Institute of Animal Reproduction and Food Research, Polish Academy of Science.

Conflicts of Interest: The authors declare no conflict of interest.

References

1. Godefroidt, T.; Ooms, N.; Pareyt, B.; Brijs, K.; Delcour, J.A. Ingredient Functionality During Foam-Type Cake Making: A Review. *Compr. Rev. Food Sci. Food Saf.* **2019**, *18*, 1550–1562. [CrossRef]
2. Psimouli, V.; Oreopoulou, V. The effect of alternative sweeteners on batter rheology and cake properties. *J. Sci. Food Agric.* **2012**, *92*, 99–105. [CrossRef]
3. Welsh, J.A.; Sharma, A.; Cunningham, S.A.; Vos, M.B. Consumption of added sugars and indicators of cardiovascular disease risk among US adolescents. *Circulation* **2011**, *123*, 249–257. [CrossRef] [PubMed]
4. Te Morenga, L.A.; Howatson, A.J.; Jones, R.M.; Mann, J. Dietary sugars and cardiometabolic risk: Systematic review and meta-analyses of randomized controlled trials of the effects on blood pressure and lipids. *Am. J. Clin. Nutr.* **2014**, *100*, 65–79. [CrossRef] [PubMed]
5. Makarem, N.; Bandera, E.V.; Nicholson, J.M.; Parekh, N. Consumption of sugars, sugary foods, and sugary beverages in relation to cancer risk: A systematic review of longitudinal studies. *Annu. Rev. Nutr.* **2018**, *38*, 17–39. [CrossRef] [PubMed]
6. Malik, V.S.; Schulze, M.B.; Hu, F.B. Intake of sugar sweetened beverages and weight gain: A systematic review. *Am. J. Clin. Nutr.* **2006**, *84*, 274–288. [CrossRef]
7. Morris, M.J.; Beilharz, J.E.; Maniam, J.; Reichelt, A.C.; Westbrook, R.F. Why is obesity such a problem in the 21st century? The intersection of palatable food, cues and reward pathways, stress, and cognition. *Neurosci. Biobehav. Rev.* **2015**, *58*, 36–45. [CrossRef]
8. Faruque, S.; Tong, J.; Lacmanovic, V.; Agbonghae, C.; Minaya, D.M.; Czaja, K. The Dose Makes the Poison: Sugar and Obesity in the United States. *Pol. J. Food Nutr. Sci.* **2019**, *69*, 219–233. [CrossRef]
9. Ng, M.; Fleming, T.; Robinson, M.; Thomson, B.; Graetz, N.; Margono, C.; Mullany, E.C.; Biryukov, S.; Abbafati, C.; Abera, S.F.; et al. Global, regional, and national prevalence of overweight and obesity in children

and adults during 1980–2013: A systematic analysis for the Global Burden of Disease Study 2013. *Lancet* **2014**, *384*, 766–781. [CrossRef]

10. Tucker, E.; Rostami, K.; Prabhakaran, S.; Al Dulaimi, D. Patients with Coeliac Disease Are Increasingly Overweight or Obese on Presentation. *J. Gastrointestin. Liver Dis.* **2012**, *21*, 11–15.

11. Mahmud, F.H.; Murray, J.A.; Kudva, Y.C.; Zinsmeister, A.R.; Dierkhising, R.A.; Lahr, B.D.; Dyck, P.J.; Kyle, R.A.; El-Youssef, M.; Burgart, L.J.; et al. Celiac Disease in Type 1 Diabetes Mellitus in a North American Community: Prevalence, Serologic Screening, and Clinical Features. *Mayo Clin. Proc.* **2005**, *80*, 1429–1434. [CrossRef] [PubMed]

12. Assor, E.; Marcon, M.A.; Hamilton, N.; Fry, M.; Cooper, T.; Mahmud, F.H. Design of a Dietary Intervention to Assess the Impact of a Gluten-Free Diet in a Population with Type 1 Diabetes and Celiac Disease. *BMC Gastroenterol.* **2015**, *15*, 181. [CrossRef] [PubMed]

13. Krupa-Kozak, U.; Lange, E. The gluten-free diet and glycaemic index in the management of coeliac disease associated with type 1 diabetes. *Food Rev. Int.* **2019**, *35*, 587–608. [CrossRef]

14. Miranda, J.; Lasa, A.; Bustamante, M.A.; Churruca, I.; Simon, E. Nutritional Differences between a Gluten-Free Diet and a Diet Containing Equivalent Products with Gluten. *Plant Foods Hum. Nutr.* **2014**, *69*, 182–187. [CrossRef] [PubMed]

15. Melini, V.; Melini, F. Gluten-Free Diet: Gaps and Needs for a Healthier Diet. *Nutrients* **2019**, *11*, 170. [CrossRef] [PubMed]

16. Kulai, T.; Rashid, M. Assessment of Nutritional Adequacy of Packaged Gluten-Free Food Products. *Can. J. Diet. Pract. Res.* **2014**, *75*, 186–190. [CrossRef]

17. Wild, D.; Robins, G.G.; Burley, V.J.; Howdle, P.D. Evidence of High Sugar Intake, and Low Fibre and Mineral Intake, in the Gluten-Free Diet. *Aliment. Pharmacol. Ther.* **2010**, *32*, 573–581. [CrossRef]

18. Vici, G.; Belli, L.; Biondi, M.; Polzonetti, V. Gluten free diet and nutrient deficiencies: A review. *Clin. Nutr.* **2016**, *35*, 1236–1241. [CrossRef]

19. Ronda, F.; Gomez, M.; Blanco, C.A.; Caballero, P.A. Effects of polyols and nondigestible oligosaccharides on the quality of sugar-free sponge cakes. *Food Chem.* **2005**, *90*, 549–555. [CrossRef]

20. Carocho, M.; Morales, P.; Ferreira, I.C.F.R. Sweeteners as food additives in the XXI century: A review of what is known, and what is to come. *Food Chem. Toxicol.* **2017**, *107*, 302–317. [CrossRef]

21. Chattopadhyay, S.; Raychaudhuri, U.; Chakraborty, R. Artificial sweeteners—A review. *J. Food Sci. Technol.* **2014**, *51*, 611–621. [CrossRef] [PubMed]

22. George, V.; Arora, S.; Wadhwa, B.K.; Singh, A.K. Analysis of multiple sweeteners and their degradation products in lassi by HPLC and HPTLC plates. *J. Food Sci. Technol.* **2010**, *47*, 408–413. [CrossRef] [PubMed]

23. Zahn, S.; Forker, A.; Krügel, L.; Rohm, H. Combined use of rebaudioside A and fibres for partial sucrose replacement in muffins. *LWT Food Sci. Technol.* **2013**, *50*, 665–670. [CrossRef]

24. Gao, J.; Brennan, M.A.; Mason, S.L.; Brennan, C.S. Effect of sugar replacement with stevianna and inulin on the texture and predictive glycaemic response of muffins. *Int. J. Food Sci. Technol.* **2016**, *51*, 1979–1987. [CrossRef]

25. Garvey, E.C.; O'Sullivan, M.G.; Kerry, J.P.; Milner, L.; Gallagher, E.; Kilcawley, K.N. Characterising the sensory quality and volatile aroma profile of clean-label sucrose reduced sponge cakes. *Food Chem.* **2020**. [CrossRef]

26. Gallagher, E.; Gormley, T.R.; Arendt, E.K. Crust and crumb characteristics of gluten free breads. *J. Food Eng.* **2003**, *56*, 153–161. [CrossRef]

27. Drabińska, N.; Zieliński, H.; Krupa-Kozak, U. Technological benefits of inulin-type fructans application in gluten-free products—A review. *Trends Food Sci. Technol.* **2016**, *56*, 149–157. [CrossRef]

28. Drabińska, N.; Rosell, C.M.; Krupa-Kozak, U. Inulin-Type Fructans Application in Gluten-Free Products: Functionality and Health Benefits. In *Bioactive Molecules in Food, Reference Series in Phytochemistry*; Mérillon, J.M., Ramawat, K., Eds.; Springer: Cham, Switzerland, 2018.

29. Mensink, M.A.; Frijlink, H.W.; Maarschalk, K.V.; Hinrichs, W.L.J. Inulin, a flexible oligosaccharide I: Review of its physicochemical characteristics. *Carbohydr. Polym.* **2015**, *130*, 405–419. [CrossRef]

30. Morais, E.C.; Cruz, A.G.; Faria, J.A.F.; Bolini, H.M.A. Prebiotic gluten-free bread: Sensory profiling and drivers of liking. *Lebensm. Wiss. Technol.* **2014**, *55*, 248–254. [CrossRef]

31. Zahn, S.; Pepke, F.; Rohm, H. Effect of inulin as a fat replacer on texture and sensory properties of muffins. *Int. J. Food Sci. Technol.* **2010**, *45*, 2531–2537. [CrossRef]

32. Krupa-Kozak, U.; Drabińska, N.; Rosell, C.M.; Fadda, C.; Anders, A.; Jeliński, T.; Ostaszyk, A. Broccoli leaf powder as an attractive by-product ingredient: Effect on batter behaviour, technological properties and sensory quality of gluten-free mini sponge cake. *Int. J. Food Sci. Technol.* **2019**, *54*, 1121–1129. [CrossRef]

33. Hsu, C.L.; Chen, W.; Weng, Y.M.; Tseng, C.Y. Chemical composition, physical properties, and antioxidant activities of yam flours as affected by different drying methods. *Food Chem.* **2003**, *83*, 85–89. [CrossRef]

34. Palou, E.; López-Malo, A.; Barbosa-Cánovas, G.V.; Welti-Chanes, J.; Swanson, B.G. Polyphenoloxidase activity and color of blanched and high hydrostatic pressure treated banana puree. *J. Food Sci.* **1999**, *64*, 42–45. [CrossRef]

35. Mokrzycki, W.S.; Tatol, M. Color difference Delta E—A survey. *Mach. Graph. Vis.* **2011**, *20*, 383–411.

36. Michalska, A.; Amigo-Benavent, M.; Zielinski, H.; del Castillo, M.D. Effect of bread making on formation of Maillard reaction products contributing to the overall antioxidant activity of rye bread. *J. Cereal Sci.* **2008**, *48*, 123–132. [CrossRef]

37. Zieliński, H.; del Castillo, M.D.; Przygodzka, M.; Ciesarova, Z.; Kukurova, K.; Zielińska, D. Changes in chemical composition and antioxidative properties of rye ginger cakes during their shelf-life. *Food Chem.* **2012**, *135*, 2965–2973. [CrossRef]

38. ISO. 8586-1: *Sensory Analysis—General Guidance for the Selection, Training and Monitoring of Assessors—Part 1: Selected Assessors*; ISO: Geneva, Switzerland, 1993.

39. Lawless, H.T.; Heymann, H. *Sensory Evaluation of Food—Principles and Practices*; Springer: New York, NY, USA, 2010.

40. ISO/DIS. 13299: *Sensory Analysis—Methodology—General Guidance for Establishing a Sensory Profile*; ISO: Geneva, Switzerland, 1998.

41. ISO. 8589: *Sensory Analysis—General Guidance for the Design of Test Rooms*; ISO: Geneva, Switzerland, 1998.

42. Rosell, C.M.; Collar, C. Effect of temperature and consistency on wheat dough performance. *Int. J. Food Sci. Technol.* **2009**, *44*, 493–502. [CrossRef]

43. Andrade, F.J.E.T.; Albuquerque, P.B.S.; Moraes, G.M.D.; Farias, M.D.P.; Teixeira-Sá, D.M.A.; Vicente, A.A.; Carneiro-da-Cunha, M.G. Influence of hydrocolloids (galactomannan and xanthan gum) on the physicochemical and sensory characteristics of gluten-free cakes based on fava beans (*Phaseolus lunatus*). *Food Funct.* **2018**, *9*, 6369–6379. [CrossRef]

44. Marco, C.; Rosell, C.M. Effect of different protein isolates and transglutaminase on rice flour properties. *J. Food Eng.* **2008**, *84*, 132–139. [CrossRef]

45. Gularte, M.A.; de la Hera, E.; Gómez, M.; Rosell, C.M. Effect of different fibers on batter and gluten-free layer cake properties. Lebensm. *Wiss. Technol.* **2012**, *48*, 209–214. [CrossRef]

46. Conforti, F.D. Cake manufacture. In *Bakery Products: Science and Technology*, 2nd ed.; Hui, Y.H., Corke, H., De Leyn, I., Nip, W., Cross, N.A., Eds.; Wiley-Blackwell: Ames, IA, USA, 2014; pp. 565–584.

47. Mondal, A.; Dutta, A.K. Bread baking—A review. *J. Food Eng.* **2008**, *86*, 465–474. [CrossRef]

48. Lostie, M.; Peczalski, R.; Andrieu, J. Lumped model for sponge cake baking during the "crust and crumb" period. *J. Food Eng.* **2004**, *65*, 281–286. [CrossRef]

49. Purlis, E.; Salvadori, V.O. Bread baking as a moving boundary problem. Part 1: Mathematical modelling. *J. Food Eng.* **2009**, *91*, 428–433. [CrossRef]

50. Dhingra, D.; Michael, M.; Rajput, H.; Patil, R.T. Dietary fibre in foods: A review. *J. Food Sci. Technol.* **2012**, *49*, 255–266. [CrossRef] [PubMed]

51. Rodríguez-García, J.; Salvador., A.; Hernando., I. Replacing Fat and Sugar with Inulin in Cakes: Bubble Size Distribution. Physical and Sensory Properties. *Food Bioprocess Technol.* **2014**, *7*, 964–974.

52. Chiavaro, E.; Vittadini, E.; Corradini, C. Physicochemical characterization and stability of inulin gels. *Eur. Food Res. Technol.* **2007**, *225*, 85–94. [CrossRef]

53. Ziobro, R.; Korus, J.; Juszczak, L.; Witczak, T. Influence of inulin on physical characteristics and staling rate of gluten-free bread. *J. Food Eng.* **2013**, *116*, 21–27. [CrossRef]

54. Starowicz, M.; Zieliński, H. How Maillard Reaction Influences Sensorial Properties (Color, Flavor and Texture) of Food Products? *Food Rev. Int.* **2019**, *35*, 707–725. [CrossRef]

55. Pérez-Quirce, S.; Collar, C.; Ronda, F. Significance of healthy viscous dietary fibres on the performance of gluten-free rice-based formulated breads. *Int. J. Food Sci. Technol.* **2014**, *49*, 1375–1382. [CrossRef]

56. Przygodzka, M.; Zieliński, H.; Ciesarová, Z.; Kukurová, K.; Lamparski, G. Effect of selected spices on chemical and sensory markers in fortified rye-buckwheat cakes. *Food Sci. Nutr.* **2016**, *4*, 651–660. [CrossRef]

Foods **2020**, *9*, 1735

57. Tamanna, N.; Mahmood, N. Food Processing and Maillard Reaction Products Effect on Human Health and Nutrition. *J. Food Chem.* **2015**. [CrossRef] [PubMed]

58. Baryłko-Pikielna, N. *Sensoryczne Badania Żywności: Podstawy, Metody, Zastosowania*; Polskie Towarzystwo Technologii Żywności; Wydawnictwo Naukowe PTTŻ: Kraków, Poland, 2009.

59. Sahi, S.S. Interfacial properties of the aqueous phases of wheat flour doughs. *J. Cereal Sci.* **1994**, *20*, 119–127. [CrossRef]

60. Meullenet, J.F.; Lyon, B.G.; Carpenter, J.A.; Lyon, C.E. Relation between sensory and instrumental texture profile attributes. *J. Sens. Stud.* **1998**, 77–93. [CrossRef]

61. Szczesniak, A.S. Texture profile analysis—Methodology interpretation clarified. *J. Food Sci.* **1995**, *6*, 60.

62. Lyon, B.G.; Champagne, E.T.; Vinyard, B.T.; Windham, W.R. Sensory and Instrumental Relationships of Texture of Cooked Rice from Selected Cultivars and Postharvest Handling Practices. *Cereal Chem.* **2000**, *77*, 64–69. [CrossRef]

Publisher's Note: MDPI stays neutral with regard to jurisdictional claims in published maps and institutional affiliations.

Article

Effect of Hydration on Gluten-Free Breads Made with Hydroxypropyl Methylcellulose in Comparison with Psyllium and Xanthan Gum

Mayara Belorio * and **Manuel Gómez**

College of Agricultural Engineering, University of Valladolid, 34004 Palencia, Spain; pallares@iaf.uva.es
* Correspondence: beloriom@gmail.com; Tel.: +34-979-108-495

Received: 28 August 2020; Accepted: 22 October 2020; Published: 26 October 2020

Abstract: The use of hydrocolloids in gluten-free breads is a strategy to improve their quality and obtain products with acceptable structural and textural properties. Hydration level (HL) optimization is important to maximize the hydrocolloids effects on dough and bread quality. This study evaluated the optimum hydration level (OHL) for gluten-free breads prepared with different starch sources (rice flour or maize starch) and hydroxypropyl methylcellulose (HPMC) in comparison with psyllium husk fibre and xanthan gum. Breads with the same final volume and the corrected hydration (CH) were evaluated. The hydration is a key factor that influences the final characteristics of gluten-free breads. Breads made with HPMC had greater dependence on the HL, especially for preparations with maize starch. Psyllium had similar behaviour to xanthan with respect to specific volume and weight loss. Breads manufactured with maize starch and HPMC had low hardness due to their great specific volume. However, in breads made with rice flour, the combined decreased hydration and similar specific volume generated a harder bread with HPMC than the use of psyllium or xanthan. Breads made with HPMC presented higher specific volume than the other hydrocolloids, however combinations among these hydrocolloids could be evaluated to improve gluten-free breads quality.

Keywords: gluten-free bread; hydration; hydroxypropyl methylcellulose; xanthan gum; psyllium

1. Introduction

Gluten plays an important role in bread formulation. The gluten network is formed by wheat proteins that with correct hydration and mechanical work, form a cohesive, extensible and elastic dough, which is able to retain the gas formed during fermentation and baking [1]. To elaborate gluten-free breads, it is necessary to resort to starches and gluten-free flours, but it is also important to replace gluten with another ingredient. However, a functionally equivalent ingredient has not yet been found that allows the full replacement of gluten. The most often used ingredients for this purpose are hydrocolloids [2–5].

Hydroxypropyl methylcellulose (HPMC) and xanthan gum are the hydrocolloids most often used as gluten substitutes in gluten-free breads, while rice flour and maize starch are the starchy ingredients most often employed in these formulations, both in scientific articles and in commercial products [6,7]. In commercial products, the use of psyllium is also prominent. In fact, Román et al. [7] indicated that 16% of all evaluated breads included psyllium as the major gluten replacer and 34% incorporated it as a secondary replacer, mixed with another main hydrocolloid. Similarly, as other hydrocolloids, psyllium is a natural fibre with important hydration and gel-forming properties [8]. It is an arabinoxylan composed by different monosaccharides and, as other hydrocolloids, it has many hydroxyl groups in its structure which increase its capacity to bind water and generate viscous solutions [9,10]. The use of psyllium has some advantages because, besides being a natural product, it is responsible for health benefits such as the regulation of glucose in diabetic disease and decreased symptoms of constipation, diarrhoea, irritable bowel syndrome and others [11]. Psyllium and xanthan gum present similar rheological behaviours, as both are responsible for weak gelling properties [12]. Nevertheless, studies about the elaboration of gluten-free breads with psyllium are scarce, and this fibre has always been studied in mixtures with other hydrocolloids such as HPMC and xanthan gum [13–15] but never as a unique gluten replacer.

Dough hydration in gluten-free bread is a fundamental aspect of final product quality. In general, it is known that the greater the hydration, the higher the specific volume of a bread, until a maximum point at which the weak structure of the dough promotes collapse during the fermentation or baking process [16,17]. However, these studies were based on doughs elaborated with HPMC, and there is little or no information about the effect of hydration on doughs made with other hydrocolloids. Generally, previous scientific studies applied the same hydration levels, despite possible changes in the formulation, or they modified hydration based on pre-proofs which were not detailed. Some authors have attempted to correct hydration by performing rheological analysis (rheometer or farinographic) [18–21]. Similarly, Ren et al. [22] evaluated the effect of hydration in psyllium and methylcellulose breads using a response surface design. However, this system had an important limitation, due to the amount of each hydrocolloid used in the analysis. These quantities did not cover a wide range of concentrations and they were the same for all hydrocolloids, so the optimal point is usually difficult to find for each case. Nevertheless, Sahagún and Gómez [23] proved that distinct gluten-free formulations achieve a maximum specific volume with differences in both hydration levels and rheological properties. Furthermore, hydration influences bread volume differently, depending on the formulation used.

Gluten-free breads prepared with the most used gluten-substitutes (HPMC or xanthan gum) were compared with breads made with psyllium in doughs with rice flour or maize starch. Therefore, the objective of this study was to evaluate the different effect of using psyllium as a gluten replacer in gluten-free breadmaking. The influence of these hydrocolloids was evaluated on dough hydration and on the specific volume of the final breads. For each case, breads with the highest specific volume were analysed in terms of crust colour and texture (hardness, springiness, cohesiveness, chewiness and resilience).

2. Materials and Methods

2.1. Bread Ingredients

Gluten-free breads were made with rice flour containing 7.54 g/100 g of protein (Molendum Ingredients SL, Zamora, Spain) or maize starch (Tereos, Syral Iberia SAU, Zaragoza, Spain). Other ingredients used were refined sunflower oil (Urzante, Navarra, Spain), sucrose (AB Azucarera Iberica, Valladolid, Spain), instant dry baker's yeast (Dosu Maya Mayacilik A.S, Istanbul, Turkey), salt (Disal, Unión Salinera de España S.A, Madrid, Spain) and tap water. Gluten replacers were hydroxypropyl methylcellulose (HPMC) (Vivapur K4M, J. Rettenmaier and Söhne, Rosenberg, Germany), xanthan gum (Industrias Roko S.A., Llanera, Asturias, Spain) and psyllium husk fibre (psyllium P95) with 80% of total fibre (14% insoluble and 66% soluble, data provided by the supplier Rettenmaier Ibérica, Barcelona, Spain) and 3.42 of water holding capacity [8].

2.2. Gluten-Free Breadmaking

A gluten-free bread recipe was composed as follows (per 100 g flour or starch): 100 g of maize starch (MS) or rice flour (RF), 6 g of sunflower oil, 5 g of sugar, 3 g of yeast powder, 1.8 g of salt and 2 g of hydrocolloid (HPMC, xanthan gum or psyllium). The amount of water was defined according to topic 2.3.

All the ingredients were mixed by using a Kitchen Aid Professional mixer (Kitchen Aid, St. Joseph, MI, USA) with a dough hook (K45DH) at 58 rpm for 1 minute, except for the dry yeast and tap water. During this minute, the water was placed in a plastic container, the dry yeast was gently laid on top of the water and it was carefully mixed with the use of a glass rod to guarantee the hydration of the whole yeast. Subsequently, the hydrated yeast was mixed (90 rpm for 8 into the dough. Portions of bread dough (150 g) were placed into aluminium pans (127 × 98 × 33 mm) previously coated with sunflower oil. The dough was fermented in a proofing chamber at 30 °C and 80% relative humidity for 60 min. The fermented doughs were baked at 190 °C for 40 minutes. The aluminium pans were removed, and the bread was allowed to cool for 60 min and placed in plastic bags, which were closed properly and stored at 20 °C for 24 h until subsequent analysis. All studied formulations were produced in duplicate.

2.3. Defining the Optimum Hydration Level of Breads

The influence of hydration level was evaluated for each bread formulation considering the use of RF, MS and the different hydrocolloids (HPMC, xanthan gum and psyllium husk fibre), similarly to a study by Sahagún and Gómez [23]. Breads were made with formulations containing 70, 80, 90, 100, 110 and 120 g/100 g of water. Their specific volumes were obtained, and those with the maximum specific volume were considered to have the optimum hydration level (OHL). The volume of all breads was measured by using a Volscan Profiler 300 (Stable Microsystems, Surrey, UK), and the specific volume was calculated as the ratio between the final volume and weight of breads, 24 h after baking. Four loaves of bread were evaluated for each formulation.

To evaluate the physical characteristics of the breads, the OHL was chosen. The amount of dough to be used in the aluminium pans was recalculated for each formulation with the aim of obtaining the same final bread volume (680 mL). However, this volume exceeded the mould capacity, and it was observed that while some doughs could grow beyond the moulds, others overgrew it and exceeded the upper edge of the pan, causing the dough drop outside the mould without increasing the bread volume. In these cases, dough hydration was reduced to the percentage at which the bread could rise without the dough dropping outside the mould.

2.4. Physical Characteristics of Breads

The weight lost during baking was calculated as the difference between the bread and dough weights divided by the dough weight.

The texture of two central slices (20 mm thick) from two breads of each formulation was evaluated. A TPA (texture profile analysis) was performed by using a TA-XT2 texture analyser (Stable Micro Systems, Godalming, UK) with a cylindrical probe 25 mm in diameter. The probe penetrated 50% of the depth of each slice, with a trigger force of 5 g and a test speed of 1 mm/s. A delay of 10 s between the first and second compressions was applied. Hardness, springiness, cohesiveness and resilience were measured.

Crumb colour was measured by using a Minolta CM-508i spectrophotometer (Minolta Co., Ltd., Osaka, Japan) with D65 as the standard illuminant and a 2° standard observer. The results were expressed in the CIE *L*a*b** colour space. Measurements were made on two central slices of two breads from each formulation (2 × 2 × 2).

2.5. Statistical Analysis

Analysis of variance (ANOVA) was performed with Statgraphics Centurion XVI software (Statpoint Technologies, Inc., Warrenton, VA, USA) to evaluate all the results obtained. The 95% confidence intervals were described by Fisher's least significant differences (LSD) test.

3. Results

3.1. Optimum Hydration Level

Specific volumes for each hydration level of gluten-free breads made with different hydrocolloids are shown in Figure 1. Among breads made with RF (Figure 1a), the specific volume of all breads increased with high hydration levels up to 100 g water/100 g flour. This increase was much larger in breads with HPMC than in those with psyllium husk fibre (PHF) or xanthan gum. In fact, breads with HPMC had a specific volume more than double of those obtained with the other hydrocolloids at 100% hydration, whereas at 70% hydration, they had similar specific volumes. The highest specific volume observed in breads elaborated with HPMC is in accordance with the results of Sabanis and Tzia [3]. Nevertheless, their study showed smaller differences than those found in this research, as breads prepared with HPMC were less hydrated than those formulated with xanthan, which reduced their volume. This is related to the water retention capacity of xanthan gum, which generates more viscous doughs. However, as shown in Figure 1, breads made with HPMC with a similar hydration of those made with xanthan gum, presented higher specific volume. The behaviour of HPMC is related to its capacity to form a thermo-reversible gel during baking, which increases the viscosity and establishes gas cell walls, providing high volume by preventing moisture loss [24]. From the 100% hydration level, the specific volume of breads with HPMC decreased, while the pattern for breads made with xanthan was a reduction at 110% and a slightly increase at 120%. However, at this hydration level the volume of the dough was reduced during fermentation, and a small increase was observed during baking. As the final volume of breads made with xanthan gum was not improved at 120% hydration, the OHL was defined as 100%, considering that at this level, the highest specific volume was obtained without the dough dropping during fermentation or baking. An increasing in the specific volume with increasing hydration up to a certain limit was also observed by Mancebo et al. [16] in breads elaborated with RF and by Sahagún and Gómez [23] in breads with MS. Both studies analysed breads with HPMC. Mancebo et al. [16] reported the specific volume of breads with optimum values of G′ and G″. Sahagún and Gómez [23] found that these rheological values depended on the bread formulation. Ziobro et al. [20] evaluated breads made with starches, guar gum and pectin, and they showed that there is a viscosity limit value at which bread volume decreases during baking. In fact, Mir et al. [2] affirmed that the internal viscosity of doughs should not be too low to avoid the release of bubbles during baking. Encina-Zelada et al. [25] also observed that the specific volume of breads made with xanthan gum or guar gum increased with increasing hydration. In this case, a relation with dough rheology was also mentioned, and it was shown that for high levels of xanthan it was necessary the addition of extra water.

In the case of PHF, doughs greater than 100% of hydration, although growing during fermentation, dropped over the edges of the moulds during baking. It is possible that the viscosity was reduced during the early stages of baking (before gelatinization) because of the increase in temperature, which promoted excessively liquid doughs with a weak structure that dropped over the edges of the mould. Thus, it was not possible to obtain properly baked rice bread containing 2% of PHF over 100% hydration (Figure 1). The specific volume of breads made with PHF was similar to those made with xanthan gum. This could be related to the rheological properties of psyllium husk fibre, which are very similar to those of xanthan gum [12].

Among breads made with MS, those with HPMC had the highest specific volume at the optimum hydration of 80% (Figure 1), and their volume gradually decreased with increasing hydration. This optimum is similar to the results obtained by Sahagún and Gómez [23] using a very similar formulation. Breads with xanthan gum presented an OHL at 110% hydration, because at 120% the volume of the dough decreased during fermentation and increased again during baking, but it was not larger than that obtained with 110%, and there was no significant difference between the two hydration levels. MS breads with PHF increased in specific volume up to 90% hydration; however, at this level, breads were completely hollow, which indicated that the dough structure was too weak and the interior matrix sunk during baking, while at the external surface, a thin crust was formed because of drying that occurred at the beginning of baking. As a result, it was not possible to measure these breads because of their weak structures. MS breads behaved similarly to RF breads, as formulations with HPMC at the OHL had nearly double the specific volume of those breads elaborated with psyllium husk fibre or xanthan gum; the differences between the latter two were small.

It is important to highlight that breads made with MS had a higher specific volume than those made with RF, considering all hydration levels and the different hydrocolloids used. All hydrocolloids (HPMC, PHF and xanthan gum) increased the final specific volume of breads by almost 50%, considering the maximum specific volume obtained for each of them. The highest specific volumes were previously found when using HPMC, comparing breads elaborated with MS to those with RF [15,26]. Martinez and Gómez [26] hypothesized that this different behaviour of RF and MS doughs could be attributed to a higher consistency of rice flour in respect to maize starch, probably due the presence of proteins. The authors also suggested another possible explanation, which is based on the presence of a protein layer that covers the starch granules of the flour, modifying the pasting behaviour and increasing the pasting temperature. With respect to the OHL, breads made with HPMC clearly had lower OHL in the presence of MS than in the presence of RF, but in the case of xanthan gum, this value was the same with MS and RF, considering that with RF the specific volume did not have significant difference when the hydration increased from 110% to 120%. In the case of PHF, it was not possible to compare values of OHL, since they were not determined by considering the maximum specific volume but because of structural problems discovered in the cases of high hydration levels. However, considering the use of psyllium husk fibre, the OHL was larger with MS (sinking of the internal structure) than RF (dough dropped outside of the mould during baking).

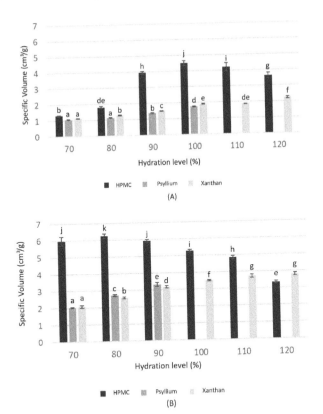

Figure 1. Variation of specific volume at different hydration levels for each gluten-free bread formulation and hydrocolloid: (**A**) rice flour (RF); (**B**) maize starch (MS). Same letters above bars means that there are no significant differences between those values ($p < 0.05$).

3.2. Gluten-Free Bread Properties

The initial idea was to produce breads with the same specific volume by changing the amount of water in the formulation, so this factor would not influence the study of the texture. It was also considered that the doughs could expand above the mould without overflowing and the hydration was reduced in the case it happened. However, the specific volume of breads made with maize starch had to be a little higher than those made with rice flour, since it was not possible to obtain such low specific volumes with MS. On the other hand, in breads made with maize starch and HPMC, it was not possible to achieve specific volumes as low as those obtained with psyllium husk fibre or xanthan gum, since reducing the hydration to a great extend generated excessively dry doughs that were difficult to handle. Table 1 shows the hydration levels used (corrected hydration, CH) and the specific volumes of breads elaborated with the CH. In the case of RF breads, it was necessary to use less hydration with the HPMC to obtain equal specific volumes of breads made with PHF and xanthan gum, according with the results obtained in the first part of this study. The CH of psyllium husk fibre and xanthan gum breads was slightly lower than the optimum found in the first part, since with the optimum hydration the doughs lost volume when they went over the edge of the mould. PHF and xanthan breads were also inferior to the optimum, due to the same problem found in RF breads, but somewhat superior to RF bread, as the doughs with these hydrations did not overflow. Regarding breads with HPMC, as it was not possible to achieve the specific volume, the hydration was corrected. In the cases in which it was not possible to equalize the specific volume, the amount of dough added in the moulds was modified to obtain breads with a similar final volume, as shown in Figure 2. Thus, all the breads had the same surface area, and this factor did not influence the weight loss after baking.

Table 1. Optimum hydration level, specific volume and weight loss for each gluten-free bread and hydrocolloid used in the formulation.

Samples	CH (%)	Specific Volume (cm³/g)	Weight Loss (g/100g)
RF HPMC	70	1.33 ± 0.01 [a]	9.33 ± 0.88 [a]
RF Psyllium	90	1.44 ± 0.02 [ab]	9.89 ± 0.18 [a]
RF Xanthan	90	1.48 ± 0.03 [b]	9.76 ± 0.18 [a]
MS HPMC	80	7.58 ± 0.04 [d]	28.20 ± 0.12 [c]
MS Psyllium	80	2.37 ± 0.08 [c]	16.60 ± 0.86 [b]
MS Xanthan	80	2.25 ± 0.08 [c]	17.50 ± 1.50 [b]

Data are expressed as means ± Standard Deviation (SD) of duplicate assays. Values with the same letter in the same column do not present significant differences ($p < 0.05$). CH: corrected hydration. RF: rice flour. HPMC: hydroxypropyl methylcellulose. MS: maize starch.

The weight loss during baking (Table 1) had no significant differences between breads made with RF. However, breads made with MS presented higher weight losses than those made with RF, and among these, those made with HPMC were the ones that presented the highest losses. In general, these weight losses are related to the volume of the loaves and the surface area, in a way that the greater is the volume, and therefore the biggest is the exchange surface, the greater will be the weight losses [16]. However, in this study, all breads had the same final volume and the same exchange surface area, so the changes in the loss of water during the baking process must be attributed to the different water absorption capacity of the ingredients used. This explains the absence of differences between breads made with PHF and xanthan, as the mixtures between starch and both ingredients have similar capacity to absorb water [8]. The higher weight loss in HPMC breads made with MS may be due to the smaller water holding capacity of this hydrocolloid [27]. The differences between starch and rice flour breads may be due to the lower water holding capacity of starches compared to flours [28], which may be related to the higher protein content of flours. In addition, a very compact structure, due to the lower specific volume of the RF breads, can also decrease the loss of water.

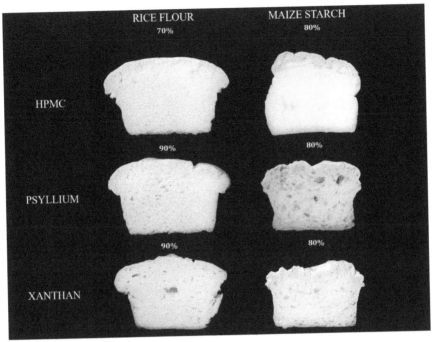

Figure 2. Crumb from maize starch or rice flour gluten-free breads and different hydrocolloids.

The results for breads texture (Table 2), show no significant differences in hardness between MS breads made with psyllium husk fibre or xanthan gum, but those made with HPMC have a considerably lower hardness. This may be due to the higher specific volume of these breads, since, in general, the higher is the specific volume, the lower is the hardness, as has been observed in previous studies with HPMC [3]. In fact, this relation between specific volume and hardness was indicated in other studies [16,26,29]. For the same reason, several studies found an increase in the hardness of gluten-free breads with the use of xanthan gum when compared with other hydrocolloids [30,31]. However, breads made with HPMC, despite of being softer, have been described as drier and with a crumblier texture [32]. Contrary to what has been observed for MS bread, in RF breads, where specific volumes were equalized, breads with HPMC presented much greater hardness than the others. This result is not observed when these products present a greater specific volume. The greater hardness of breads with HPMC found in this study may be attributed to the gels reverting to a weakly entangled form upon cooling, which increased crumb firmness after baking [29,33]. Nevertheless, despite the higher specific volume of breads elaborated with MS, RF breads made with PHF showed similar hardness and those with xanthan gum were less hard than MS breads. These differences can be explained by the distinct effect of xanthan gum on the pasting properties of starches, which includes the retrogradation phenomenon [28]. Thus, it seems that maize starch generates harder breads than rice flour. This is in accordance with the findings of Mancebo et al. [34], who reported that the texture of breads made with starch was harder than that of breads made with rice flour. RF breads made with xanthan and psyllium fiber husk showed higher springiness, cohesiveness and resilience in respect to RF-HPMC breads, possibly due to the same reason, regarding the changes in the HPMC gels after baking. On the contrary, in MS breads these changes were not found, as they were counteracted by the effect of the specific volume of the breads. As regards the PHF and xanthan breads, they showed almost similar values, and only slightly significant differences could be seen in MS breads, where PHF breads showed greater cohesiveness and resilience than xanthan breads.

Table 2. Texture parameters of gluten-free breads made with RF or MS for each hydrocolloid.

Samples	Hardness (N)	Springiness	Cohesiveness	Resilience
RF HPMC	42.44 ± 0.21 [d]	0.796 ± 0.004 [a]	0.656 ± 0.023 [ab]	0.383 ± 0.009 [a]
RF Psyllium	14.98 ± 0.60 [c]	0.891 ± 0.025 [b]	0.748 ± 0.037 [c]	0.479 ± 0.041 [bc]
RF Xanthan	9.04 ± 3.00 [b]	0.922 ± 0.043 [bc]	0.807 ± 0.024 [c]	0.501 ± 0.013 [bc]
MS HPMC	1.44 ± 0.12 [a]	1.011 ± 0.023 [d]	0.754 ± 0.030 [c]	0.493 ± 0.034 [bc]
MS Psyllium	19.51 ± 3.40 [c]	0.974 ± 0.004 [cd]	0.733 ± 0.037 [bc]	0.550 ± 0.052 [c]
MS Xanthan	19.58 ± 1.55 [c]	0.964 ± 0.002 [cd]	0.606 ± 0.037 [a]	0.420 ± 0.047 [ab]

Data are expressed as means ± SD of duplicate assays. Values with the same letter in the same column are not significantly different ($p < 0.05$). RF: rice flour. MS: maize starch.

Comparing the crust colour (Table 3), breads made with xanthan gum were darker (small values of *L**) than breads with HPMC, and no significant differences were observed between breads made with HPMC and PHF. Neither were significant differences observed between RF breads elaborated with PHF or xanthan gum. Values of *a** and *b** had small significant differences, and no clear tendency was observed. However, breads made with xanthan gum had the largest values of *a** in preparations containing RF and the smallest values of *b** among those made with MS. Breads containing HPMC presented the highest values of *a** and *b** among preparations with MS. The crust colour of breads is related to the Maillard reaction, which occurs between amino acids and reducing sugars, as well as sugar caramelization [35]. Differences in sugar content and amino acids should not exist between breads elaborated with different hydrocolloids, since the other ingredients, which are the responsible for sugar and amino acid contents, did not change in the formulation. However, water activity can vary depending on the hydrocolloid and hydration of the doughs and this can affect Maillard reactions favouring the mobility of reactants [36]. In fact, Sabanis and Tzia [3] also found significant differences among the crust colours of breads made with distinct hydrocolloids.

Table 3. Crust colour parameters of gluten-free breads.

Samples	*L**	*a**	*b**
RF HPMC	81.68 ± 3.05 [c]	1.64 ± 0.25 [bc]	17.15 ± 0.68 [bc]
RF Psyllium	79.92 ± 4.67 [bc]	1.23 ± 0.39 [b]	15.36 ± 0.18 [b]
RF Xanthan	75.05 ± 0.83 [ab]	4.48 ± 0.08 [d]	20.25 ± 1.22 [c]
MS HPMC	82.09 ± 0.04 [c]	2.64 ± 0.14 [c]	19.32 ± 0.22 [c]
MS Psyllium	86.20 ± 2.13 [c]	-0.05 ± 1.07 [a]	14.56 ± 2.64 [b]
MS Xanthan	71.26 ± 1.92 [a]	0.06 ± 0.09 [a]	9.72 ± 1.93 [a]

Data are expressed as means ± SD of duplicate assays. Values with the same letter in the same column are not significantly different ($p < 0.05$). RF: rice flour. MS: maize starch.

4. Conclusions

In general, breads manufactured with HPMC and maize starch showed higher specific volumes than preparations with other hydrocolloids or rice flour. Nevertheless, the degree of hydration of the dough can change these results. The hydration effect is much more evident in breads prepared with HPMC than in those made with psyllium or xanthan gum. Thus, optimization of hydration is fundamental when different gluten-free breads are evaluated. After obtaining similar specific volume due to the corrected hydration, breads made with RF and HPMC were harder and less cohesive than breads made with psyllium or xanthan gum. It was not possible to obtain breads with the same specific volume because of great differences among the different hydrocolloids. Thus, breads made with HPMC and MS presented higher specific volume and lower hardness, but they presented a high weight loss during baking. Psyllium behaved similarly as xanthan gum, both with rice flour and maize starch. It is important to optimize the hydration for all gluten-free bread formulations made by using RF or MS with PHF, xanthan gum or HPMC. The optimum hydration allows to achieve high specific volumes, however in HPMC breads with RF and, especially, with MS it is possible to obtain specific volumes higher than breads with PHF or xanthan gum. Thought, it is necessary to study the combination between these hydrocolloids and starch sources to optimize the texture of final breads.

Author Contributions: Conceptualization, M.G.; methodology, M.G. and M.B.; validation, M.G.; formal analysis, M.B.; investigation, M.B.; data curation, M.G. and M.B.; writing—original draft preparation, M.B.; writing—review and editing, M.B. and M.G.; visualization, M.G.; supervision, M.G.; project administration, M.G.; funding acquisition, M.G. All authors have read and agreed to the published version of the manuscript.

Funding: This research was funded by European Regional Development Fund (0612_TRANS_CO_LAB_2_P).

Acknowledgments: This work was supported by the European Regional Development Fund through the project TRANSCOLAB (0612_TRANS_CO_LAB_2_P). The authors are also grateful to Molendum Ingredients for supplying the rice flour and Rettenmaier Iberica for supplying the psyllium and HPMC.

Conflicts of Interest: The authors declare no conflict of interest. The funders had no role in the design of the study; in the collection, analyses, or interpretation of data; in the writing of the manuscript, or in the decision to publish the results.

References

1. Delcour, J.A.; Joye, I.J.; Pareyt, B.; Wilderjans, E.; Brijs, K.; Lagrain, B. Wheat gluten functionality as a quality determinant in cereal-based food products. *Annu. Rev. Food Sci. Technol.* **2012**, *3*, 469–492. [CrossRef] [PubMed]

2. Mir, S.A.; Shah, M.A.; Naik, H.R.; Zargar, I.A. Influence of hydrocolloids on dough handling and technological properties of gluten-free breads. *Trends Food Sci. Technol.* **2016**, *51*, 49–57. [CrossRef]

3. Sabanis, D.; Tzia, C. Effect of hydrocolloids on selected properties of gluten-free dough and bread. *Food Sci. Technol. Int.* **2011**, *17*, 279–291. [CrossRef] [PubMed]

4. Sciarini, L.S.; Ribotta, P.D.; León, A.E.; Pérez, G.T. Effect of hydrocolloids on gluten-free batter properties and bread quality. *Int. J. Food Sci. Technol.* **2010**, *45*, 2306–2312. [CrossRef]

5. Anton, A.A.; Artfield, S.D. Hydrocolloids in gluten-free breads: A review. *Int. J. Food Sci. Nutr.* **2008**, *59*, 11–23. [CrossRef] [PubMed]

6. Masure, H.G.; Fierens, E.; Delcour, J.A. Current and forward looking experimental approaches in gluten-free bread making research. *J. Cereal Sci.* **2016**, *67*, 92–111. [CrossRef]

7. Román, L.; Belorio, M.; Gomez, M. Gluten-free breads: The gap between research and commercial reality. *Compr. Rev. Food Sci. Food Saf.* **2018**, *18*, 690–702. [CrossRef]

8. Belorio, M.; Marcondes, G.; Goméz, M. Influence of psyllium versus xanthan gum in starch properties. *Food Hydrocoll.* **2020**, *105*, 105843. [CrossRef]

9. Zhang, J.; Wen, C.; Zhang, H.; Duan, Y. Review of isolation, structural properties, chain conformation, and bioactivities of psyllium polysaccharides. *Int. J. Biol. Macromol.* **2019**, *139*, 409–420. [CrossRef]

10. Saha, D.; Bhattacharya, S. Hydrocolloids as thickening and gelling agents in food: A critical review. *J. Food Sci. Technol.* **2010**, *47*, 587–597. [CrossRef]

11. Singh, B. Psyllium as therapeutic and drug delivery agent. *Int. J. Pharm* **2007**, *334*, 1–14. [CrossRef] [PubMed]

12. Haque, A.; Richardson, R.K.; Morris, E.R.; Dea, I.C.M. Xanthan-like "weak gel" rheology from dispersions of ispaghula seed husk. *Carbohydr. Polym.* **1993**, *22*, 223–232. [CrossRef]

13. Cappa, C.; Lucisano, M.; Mariotti, M. Influence of psyllium, sugar beet fibre and water on gluten-free dough properties and bread quality. *Carbohydr. Polym.* **2013**, *98*, 1657–1666. [CrossRef]

14. Haque, A.; Morris, E.R. Combined use of ispaghula and HPMC to replace or augment gluten in breadmaking. *Food Res. Int.* **1994**, *21*, 379–393. [CrossRef]

15. Mancebo, C.M.; San Miguel, M.Á.; Martínez, M.M.; Gómez, M. Optimisation of rheological properties of gluten-free doughs with HPMC, psyllium and different levels of water. *J. Cereal Sci.* **2015**, *61*, 8–15. [CrossRef]

16. Mancebo, C.M.; Martínez, M.M.; Merino, C.; de la Hera, E.; Gómez, M. Effect of oil and shortening in rice bread quality: Relationship between dough rheology and quality characteristics. *J. Texture Stu* **2017**, *48*, 597–606. [CrossRef]

17. McCarthy, D.F.; Gallagher, E.; Gormley, T.R.; Schober, T.J.; Arendt, E.K. Application of response surface methodology in the development of gluten-free bread. *Cereal. Chem.* **2005**, *82*, 609–615. [CrossRef]

18. Martínez, M.M.; Oliete, B.; Román, L.; Gómez, M. Influence of the addition of extruded flours on rice bread quality. *J. Food Qual.* **2014**, *37*, 83–94. [CrossRef]

19. Nunes, M.H.B.; Ryan, L.A.M.; Arendt, E.K. Effect of low lactose dairy powder addition on the properties of gluten-free batters and bread quality. *Eur. Food Res. Technol.* **2009**, *229*, 31–41. [CrossRef]
20. Ziobro, R.; Juszczak, L.; Witczak, M.; Korus, J. Non-gluten proteins as structure forming agents in gluten free bread. *J. Food Sci. Technol.* **2016**, *53*, 571–580. [CrossRef]
21. Ziobro, R.; Witczak, T.; Juszczak, L.; Korus, J. Supplementation of gluten-free bread with non-gluten proteins. Effect on dough rheological properties and bread characteristic. *Food Hydrocoll.* **2013**, *32*, 213–220. [CrossRef]
22. Ren, Y.; Linter, B.R.; Linforth, R.; Foster, T.J. A comprehensive investigation of gluten free bread dough rheology, proving and baking performance and bread qualities by response surface design and principal component analysis. *Food Funct.* **2020**. [CrossRef] [PubMed]
23. Sahagún, M.; Gómez, M. Assessing influence of protein source on characteristics of gluten-free breads optimising their hydration level. *Food Bioproc. Technol.* **2018**, *11*, 1686–1694. [CrossRef]
24. Crockett, R.; Ie, P.; Vodovotz, Y. How do xanthan and hydroxypropyl methylcellulose individually affect the physicochemical properties in a model gluten-free dough. *J. Food Sci.* **2011**, *76*, 274–282. [CrossRef]
25. Encina-Zelada, C.R.; Cadavez, V.; Monteiro, F.; Teixeira, J.A.; Gonzales-Barron, U. Physicochemical and textural quality attributes of gluten-free bread formulated with guar gum. *Eur. Food Res. Technol.* **2019**, *245*, 443–458. [CrossRef]
26. Martínez, M.M.; Gómez, M. Rheological and microstructural evolution of the most common gluten-free flours and starches during bread fermentation and baking. *J. Food Eng.* **2017**, *197*, 78–86. [CrossRef]
27. Horstmann, S.W.; Axel, C.; Arendt, E.K. Water absorption as a prediction tool for the application of hydrocolloids in potato starch-based bread. *Food Hydrocoll.* **2018**, *81*, 129–138. [CrossRef]
28. Matia-Merino, L.; Prieto, M.; Román, L.; Gómez, M. The impact of basil seed gum on native and pregelatinized corn flour and starch gel properties. *Food Hydrocoll.* **2019**, *89*, 122–130. [CrossRef]
29. Gallagher, E.; Gormley, T.R.; Arendt, E.K. Crust and crumb characteristics of gluten free breads. *J. Food Eng.* **2003**, *56*, 153–161. [CrossRef]
30. Lazaridou, A.; Duta, D.; Papageorgiou, M.; Belc, N.; Biliaderis, C.G. Effects of hydrocolloids on dough rheology and bread quality parameters in gluten-free formulations. *J. Food Eng.* **2007**, *79*, 1033–1047. [CrossRef]
31. Schober, T.J.; Bean, S.R.; Boyle, D.L. Gluten-free sorghum bread improved by sourdough fermentation: Biochemical, rheological, and microstructural background. *J. Agric. Food Chem.* **2007**, *55*, 5137–5146. [CrossRef]
32. Liu, X.; Mu, T.; Sun, H.; Zhang, M.; Chen, J.; Laure, M. Influence of different hydrocolloids on dough thermo-mechanical properties and in vitro starch digestibility of gluten-free steamed bread based on potato flour. *Food Chem.* **2018**, *239*, 1064–1074. [CrossRef]
33. Grover, J.A. Methylcellulose (MC) and hydroxypropyl methylcellulose (HPMC). In *Food Hydrocolloids*; CRC Press: Boca Raton, FL, USA, 1982.
34. Mancebo, C.M.; Merino, C.; Martinez, M.M.; Gomez, M. Mixture design of rice flour, maize starch and wheat starch for optimization of gluten free bread quality. *J. Food Sci. Technol.* **2015**, *52*, 6323–6333. [CrossRef]
35. Purlis, E.; Salvadori, V.O. Modelling the browning of bread during baking. *Food Res. Int.* **2009**, *42*, 865–870. [CrossRef]
36. Gonzales, A.S.P.; Naranjo, G.B.; Leiva, G.E.; Malec, L.S. Maillard reaction kinetics in milk powder: Effect of water activity at mild temperatures. *Int. Dairy J.* **2010**, *20*, 40–45. [CrossRef]

Publisher's Note: MDPI stays neutral with regard to jurisdictional claims in published maps and institutional affiliations.

Article

Can Manipulation of Durum Wheat Amylose Content Reduce the Glycaemic Index of Spaghetti?

Mike Sissons [1,*], Francesco Sestili [2], Ermelinda Botticella [2], Stefania Masci [2] and Domenico Lafiandra [2,*]

1   NSW Department of Primary Industries, Tamworth 2340, Australia
2   Department of Agriculture and Forest Sciences, University of Tuscia, 01100 Viterbo, Italy; francescosestili@unitus.it (F.S.); e.botticella@unitus.it (E.B.); masci@unitus.it (S.M.)
*   Correspondence: mike.sissons@dpi.nsw.gov.au (M.S.); lafiandr@unitus.it (D.L.)

check for
updates

Received: 8 May 2020; Accepted: 24 May 2020; Published: 28 May 2020

Abstract: Resistant starch (RS) in foods has positive benefits for potentially alleviating lifestyle diseases. RS is correlated positively with starch amylose content. This study aimed to see what level of amylose in durum wheat is needed to lower pasta GI. The silencing of starch synthases IIa (SSIIa) and starch branching enzymes IIa (SBEIIa), key genes involved in starch biosynthesis, in durum wheat cultivar Svevo was performed and spaghetti was prepared and evaluated. The SSIIa and SBEIIa mutants have a 28% and 74% increase in amylose and a 2.8- and 35-fold increase in RS, respectively. Cooked pasta was softer, with higher cooking loss but lower stickiness compared to Svevo spaghetti, and with acceptable appearance and colour. In vitro starch digestion extent (area under the digestion curve) was decreased in both mutants, but much more in SBEIIa, while in vivo GI was only significantly reduced from 50 to 38 in SBEIIa. This is the first study of the glycaemic response of spaghetti prepared from SBEIIa and SSIIa durum wheat mutants. Overall pasta quality was acceptable in both mutants but the SBEIIa mutation provides a clear glycaemic benefit and would be much more appealing than wholemeal spaghetti. We suggest a minimum RS content in spaghetti of ~7% is needed to lower GI which corresponded to an amylose content of ~58%.

Keywords: durum wheat; pasta; glycaemic index; high amylose; resistant starch

1. Introduction

Food products made from wheat are and continue to be a key source of human nutrition and pleasure to civilisations around the world. While bread products dominate, noodles and pasta, which are convenient, inexpensive, and easy to cook, continue to enjoy popularity across the globe. Pasta is prepared mostly from durum wheat semolina (*Triticum turgidum* subsp. *durum*) with a small rise in the consumption of wholegrain/wholemeal and bran containing pasta occurring over the last decade [1]. Regular pasta, which is made from semolina, is not an ideal source of dietary fibre as most has been removed during the milling of the grain. There is good evidence that regular consumption of wholegrain cereals offers a reduced risk of certain diseases like type 2 diabetes, cardiovascular disease, and certain cancers [2,3] and attempts are being made by the cereal industry to get this message out with mixed success across the world. Despite this knowledge, the daily intake of dietary fibre falls well short of daily recommendations, with more than 90% of the population of the USA, for example, not meeting target levels [4]. This is because the vast majority of wheat-based food is made from refined flour where the grain outer layers, which contain most of the fibre, have been removed in milling. However, such foods are preferred by consumers for their taste, sensory quality, and appearance, so refined wheat products, by default, constitute a major, although poor, source of fibre [5]. The key would be to modify the nutritional value of the refined wheat products, while largely retaining the same sensory qualities and consumer appeal. This is a better strategy than adding ingredients to improve fibre content like bran, wholegrain meal, legumes, gums, etc. In pasta, a meta-analysis of 66 studies where pasta was fortified to improve its nutritional value showed that enrichment levels up to only 10% can be tolerated before sensory properties are compromised, which will limit potential fibre improvement by this route [6].

Resistant starch (RS), which is that fraction of dietary fibre that escapes digestion and absorption in the upper gastrointestinal tract and flows to the large intestine, serving as a substrate for resident bacteria, has the potential to improve dietary fibre in foods. RS has been found to have many benefits in foods such as lowering glycaemic index, promoting satiety, prebiotic, hypocholesterolaemic, and more [7,8]. Several commercial sources of RS are available (Hi-maize, CrystaLean®, Novelose®, Amylomaize VII and others) to increase the dietary fibre (DF) of foods [8]. These have several advantages over conventional fibres being white, with bland flavour and of fine particle size, low calorie content with total dietary fibre (TDF) over 20%, absorb little water and can be more stable in food processing. Commercial sources of RS have been added to pasta and up to 10–20% can be incorporated, depending on the source, with some impact on sensory appeal while increasing fibre among other benefits [9–12]. However, an alternative to adding commercially synthesised RS is to elevate the natural levels of RS in the wheat, which are likely to be less expensive and more acceptable to consumers, being "natural" sources of high fibre.

As the increase of amylose in the grain has been associated with increased RS [13,14] and because of the association with reduced starch digestibility and increased fibre with many potential applications in food, the development of high amylose cereals was stimulated. In the last fifteen years, wheat scientists, targeting starch biosynthesis by plant genetics tools, have developed new lines in which amylose/amylopectin ratio results have been almost reversed compared to the normal starch [15–20]. In the durum wheat Svevo, the transgenic silencing of a single key enzyme (starch branching enzyme IIa-SBEIIa) increased amylose up to 75–80% vs. the 25–30% in the control [19]. SBEIIa is a transglycosylase enzyme that catalyzes the formation of α-1,6 branches within amylopectin. Hazard et al. [21] described a durum wheat line in which mutating the gene coding the same enzyme, resulted in a modest increase of amylose content in the starch. In one of the durum wheat lines, used in this work, the gene SBEIIa was knocked-down by a successful non-transgenic technology named TILLING [19]. TILLING is a reverse genetic approach that combines the use of mutagenesis (to increase the genetic variability) and molecular high throughput techniques (to identify mutations in target genes). The semolina of the new Svevo derived line (Svevo SBEIIa) reached 55% in amylose (on total starch) and resistant starch was found to be increased by up to 6.5%. Several authors have reported a significant increase in amylose consequent to the suppression of another enzyme, the starch synthase IIa (SSIIa), essential for the elongation of amylopectin chains [22,23]. In the SSIIa null wheats, the composition of the whole seed is altered [22]. Here, the modest increase in amylose is accompanied with a severe decrease in starch content and a large increase in total fiber. These characteristics open interesting possibilities for the production of new healthy foods [24].

The purpose of this work was to assess the pasta quality and glycaemic index of two novel, non-transgenic durum wheats with elevated levels of amylose in their endosperm with a view to showing the potential for reducing the glycaemic index of pasta while maintaining acceptable pasta technological properties.

2. Materials and Methods

2.1. Plant Materials

Durum wheat lines Svevo SSIIa and Svevo SBEIIa (referred to as SSIIa and SBEIIa) were previously produced [19,22]. The two lines along with the control (durum wheat cv. Svevo) were grown in open field at the Experimental Farm of the University of Tuscia, located in Viterbo, Italy (lat. 42°26′ N, long. 12°04′ E, altitude 310 m a.s.l.) in two different seasons: Svevo SSIIa along with Svevo in the season

2015–2016, whereas Svevo SBEIIa along with Svevo in the season 2016–2017. Nitrogen fertilization (180 kg ha^{-1}) was split into three applications: the first was given before sowing as di-ammonium phosphate (20% of total N applied), the second when the first node was detectable above ground as urea (50% of total N), and the third 25 days later as ammonium nitrate (30% of total N). Weather data show a typical pattern at this location (Supplementary Table S1).

2.2. Sample Preparation and Analytical Methods

Wheat was cleaned, conditioned to a water content of about 16.5% and left to moisten overnight. Standard milling was performed in a Buhler MLU 202 mill (Buhler, Utzwil, Switzerland) with three breaking and three sizing passages [25]. Semolina protein was determined by Dumas combustion using a Leco TruMax CN combustion nitrogen analyser (Leco Corp. St. Joseph, MI, USA) calibrated with sulfamethazine [26]. Semolina moisture was determined by the approved Method 44–15A [25]. Swelling power was measured as described elsewhere [27] in duplicate. The amylose content of the semolina, and resistant starch and total starch content of ground pasta (coffee grinder, sieved across a 250 μm screen) were assayed in duplicate using Megazyme kits (Deltagen Australia, Melbourne, Australia). Starch gelatinization parameters (enthalpy, onset, peak and end temperature) were performed on isolated starch as described previously [28] except the temperature was ramped up to 120 °C at 10 °C/min. Flour water absorption, adjusted to 14% mb (FWA, 14% mb) was determined using a MicroDoughLAB (Perten Instruments, Australia) fitted with a 4-g bowl, mixing at 120 rpm to target peak 650 FU in duplicate [25]. The particle size distribution of the semolina was measured using a vibratory sieve shaker (Fritch, Analysette 3 sparatan, Germany) adjusting the amplitude to 2.0 mm with a run time of 3 min using screens of apertures 500, 425, 315, 250, and 180 μm and the amount retained on each screen was collected and weighed.

2.3. Pasta Preparation and Evaluation

Spaghetti was prepared as previously described [29] but with adjustment to water added to make the dough based on the water absorption of the semolina to account for the higher water absorption of high amylose flours [30]. For 1 kg of semolina, the amount of water added for Svevo was 290 mL; SSIIa 333 mL and SBEIIa 349 mL. Dough blends were prepared in a premixing chamber for 15 min then the dough transferred to a pasta extruder fitted with a 1.82 mm spaghetti die (Appar Laboratorio, Rome). Wet spaghetti was transferred to a drying cabinet (TEC 2604, Thermoline Scientific Equipment, Smithfield, Australia) and dried using drying program up to maximum 65 °C. Dried pasta was stored in sealed plastic bags at room temperature until required for analysis.

All pasta samples were cooked to their fully cooked time (FCT), the time taken for the central starch core to disappear [25] and assessed for texture (cooked firmness, overcooking tolerance (100 × [firmness at FCT-firmness at FCT plus 10 min overcooking/firmness at FCT], stickiness), cooking loss and water absorption as described previously [31]. For firmness and overcooking tolerance, 12 replicate tests were performed per sample, for stickiness, a minimum of four replicate analyses per sample, and for cooking loss and water absorption of pasta, duplicate analyses were collected. The colour of uncooked 7 cm spaghetti strands aligned to minimize air spaces enough to cover the Minolta Chroma meter CR-410 detector (Biolab Australia, Sydney) was performed with a minimum of 4 replicate readings calibrated with white tile supplied by manufacturer. Measurements were L* (brightness, 100 = white; 0 = black), a* (positive value is redness and negative value is greenness), and b* (positive value, yellowness; negative value, blueness). A commercial sample of wholemeal spaghetti was used for comparison (15% protein, 62% carbohydrate, 3.5% fat and 9% dietary fibre).

2.4. In Vitro Starch Digestion of Pasta

Starch digestion of the samples was determined based on previous work [32]. Six pasta strands (typically 35–50 mm in length) for each sample were cooked in 36 mL of RO water to their FCT then cooled in water and trimmed to ~5 mm length. About 9–12 pieces were added (to standardize digestions 90 mg of starch was subject to digestion) to three 100 mL conical flasks (2 replicates samples and one control—no enzymes added) to which 6 mL of pre-heated RO water was added and 5 mL of pepsin solution (1 mg/mL in 0.02 M HCl) except for control with 0.02 N HCL added. Flasks were incubated with shaking at 1400 rpm for 30 min in a water bath held at 37 °C. To terminate the reaction, 5 mL of 0.2 M sodium acetate buffer (pH 6.0) was added to each flask followed by addition of 5 mL α amylase/amyloglucosidase (AA/AMG) solution and 5 mL of buffer to controls then incubated for 360 min at 37 °C. During the incubation, at intervals, a 0.1 mL aliquot was removed from the reaction mixture and mixed with 0.9 mL ethanol (to terminate the enzyme reaction). This mixture was assayed for glucose using the Megazyme GOPOD reagent kit as per instructions. Absorbance at 510 nm was recorded using a UV mini −1240 Spectrophotometer (Shimadzu). Glucose content (mg/mL) = corrected sample absorbance (test sample absorbance–control absorbance)/absorbance glucose standard.

The starch digested (%) was calculated as:

$$\text{Glucose content} \times 10 \times 21 \times (162/180) \times (100/90)$$

where 10 is the dilution factor (0.1 mL of reaction mix added to 0.9 mL ethanol), 21 the dilution factor (1 mL to 21 mL reaction mix), 162/180 the molecular weight ratio when converting from starch to glucose, 90 the quantity of starch present in reaction mix in mg, and 100 to convert to %.

2.5. In Vivo Glycaemic Index (GI) Testing

This measurement was contracted to SUGiRS laboratory (Sydney University glycaemic index research services, https://www.glycemicindex.com/testing_research.php). In brief, 10 healthy subjects, non-smoking, aged 19–46 years with an average body mass index of 21.5–22.1 kg/m^2 within the healthy range, were selected for the study on two separate occasions (samples Svevo 2016 and SSIIa pasta tested in 2017 and Svevo 2017 and SBEIIa pasta tested in 2019). The reference food and two pasta samples were served to participants (who had fasted overnight) in fixed test portions containing 50 g of available carbohydrate. Pure glucose (Glucodin™ powder, Valeant Pharmaceuticals, NSW) dissolved in water was used as a reference food and was consumed by each participant on three separate occasions, but participants only consumed the test pasta on one occasion. Each test portion of pasta was prepared shortly before being required by weighing appropriate amount of dry pasta and cooking in boiling water to FCT, drained, rinsed under cold water and then served to a participant with a glass of 250 mL of water. Participants were required to consume all food and fluid served.

A fasting blood sample is collected and then the test or reference sample consumed after which additional blood samples are collected at regular intervals over the next 2-h period. The same procedure is repeated in the same group of people on another day after they have consumed a portion of the reference glucose. The glucose was consumed on the first, third and fifth test sessions and the pasta samples were consumed in random order in between. The night before each test session, participants ate a regular evening meal based on carbohydrate-rich food and then fasted for at least 10 h overnight. Blood samples collected were centrifuged for 45 s immediately after collection and the plasma layer removed and stored at −20 °C for later analysis. Plasma glucose was measured in duplicate using a glucose hexokinase assay and clinical chemistry analyser. A GI value for the test pasta is calculated by expressing the 2 h blood glucose area under the curve for the test pasta as a percentage of the area produced by the glucose (GI = 100).

2.6. Statistical Methods

Data were analysed using the statistical programme GenStat version 17.1.0.14713 with a generalised linear model and the means were tested for significant differences by the least significant difference statistic (LSD), $p < 0.05$. Data were checked for normality and a Pearson correlation analysis was performed to examine the relationships between measured parameters.

3. Results and Discussion

3.1. Impact of Elevated Amylose on Pasta Technological Properties

Given that the high amylose samples were developed and field trialled in different years, Svevo as control was grown in both the 2016 and 2017 seasons to be control for the SSIIa and SBEIIa, respectively. Typical amylose contents of durum wheat range from 26% to 32% [33] but this can vary with seasonal conditions with Svevo amylose in this range (Table 1). As previously demonstrated, the elimination of SSIIa and SBEIIa in durum wheat by mutations created in the *SSIIa* and *SBEIIa* genes results in seeds with a higher amylose content compared to the wild type ones [19,22,34] as we obtained (Table 1), with amylose in SSIIa increased by ~29% and in SBEIIa by ~72% compared to Svevo. Similar results were obtained in bread wheat [20,35]. The elimination of starch biosynthetic enzymes can cause a reduction in starch synthesis and reduced seed weight and the total starch in the pasta of the high amylose (HA) genotypes (SSIIa and SBEIIa) was decreased significantly (Table 2). The protein content of the Svevo semolina differed only slightly across the two growing seasons while HA genotypes showed higher protein content than their corresponding Svevo (Table 1) probably due to the starch reduction and smaller grains (data not shown). This would explain the higher grain protein content of the HA lines as observed in other HA material [36]. Botticella et al. [23] reported a smaller grain weight, 25–45% reduction in starch content and concomitant increase in protein in the genotype Svevo SSIIa. Hogg et al. [35] also reported higher grain protein, lower starch content and kernel weight, and lower swelling power in their SSIIa triple mutant, consistent with our results for Svevo SSIIa.

It is known that the ratio of amylose to amylopectin affects the water absorption of flour [30,37] and this needs to be determined to optimise the water addition for pasta preparation. Both HA semolina's have a much higher water absorption than Svevo (Table 1) with SBEIIa being a few percent higher than SSIIa. A similar behaviour was observed with waxy wheat flour [30], that attributed the increase in water absorption to the higher dietary fiber content of HA starch. While total fibre was not measured in the SSIIa genotype, it was significantly higher in SBEIIa compared to Svevo (Table 1). However, the increase in the resistant starch content of the HA pasta was significantly higher than Svevo genotypes (Table 2) supporting the increased fibre content of the HA genotypes as RS is a fibre. The SBEIIa has a much higher RS content than SSIIa pasta with both having a 35- and 2.8-fold increase above Svevo, respectively. The semolina swelling power of both HA lines were lower than Svevo (Table 1). It is known that this parameter is negatively correlated with the amylose content because amylopectin is responsible for the starch swelling and amylose acts as a diluent [38,39]. These changes are expected to have an impact on pasta texture.

Pasta in the form of spaghetti, a common shape consumed worldwide, was evaluated for its colour, cooked texture, and cooking quality parameters, which are important to consumers of pasta. Pasta cooking quality can be assessed for features typical of good quality pasta which should be firm and resilient, with minimal cooking loss and surface stickiness and increase in volume after cooking to provide an acceptable mouthfeel. The yellowness of pasta is of aesthetic importance for consumer acceptance and marketing whereas redness and brownness are considered undesirable [40]. The appearance of the pasta (Figure 1) shows both HA pastas have a very similar appearance to the Svevo. SBEIIa was duller while SSIIa was closer to Svevo in visual colour but a little darker (duller). In contrast, a commercial wholemeal pasta example appears much darker, red and more grainy.

Table 1. Semolina properties of HA and control genotypes.

Sample	SP	Amylose (%)	Protein (11%mb) [a]	Ash (14%mb)	FWA (14%mb)	TDF (%)	TIF (%)	TSF (%)	Granularity (g/100 g)					
									500 μm	425 μm	315 μm	250 μm	180 μm	<180 μm
Svevo2016	8.93 ± 0.61 [a]	34.0 ± 1.7 [a]	12.9	0.70 ± 0.049	59.4 ± 0.14 [a]	nd	nd	nd	0.07	4.1	33.8	27.5	18.6	15.3
Svevo SSIIa	6.62 ± 0.69 [b]	43.5 ± 1.5 [b]	14.2	0.82 ± 0.014	74.3 ± 0.28 [b]	nd	nd	nd	0.01	2.0	35.3	39.1	17.8	5.4
Svevo2017	10.56 ± 0.59 [a]	33.3 ± 1.3 [c]	12.8	0.53 ± 0.012	57.0 ± 0.14 [c]	2.5 [a]	1.4 [a]	1.1 [a]	0.01	0.06	43.2	36.1	17.4	2.7
Svevo SBEIIa	5.14 ± 0.45 [b]	57.8 ± 1.5 [d]	15.4	0.81 ± 0.042	77.2 ± 0.21 [d]	5.7 [b]	3.8 [b]	1.9 [b]	0.02	0.02	44.3	34.1	17.9	3.3

SP = swelling power; FWA = Flour Water Absorption; TDF = total dietary fibre; TIF = total insoluble fibre; TSF = total soluble fibre. Numbers in the same column with different superscript letters indicate significant differences (*p* < 0.05). [a] Measurement was done in commercial lab and precision of measurement for protein is with cv 1.3%. nd = not determined.

Table 2. Pasta properties of HA and control genotypes.

Sample	Field Season	Dry Pasta				
		DP-L *	DP-a *	DP-b *	RS% (dm)	TS% (dm)
Svevo	2016	70.12 ± 0.30 [a]	0.29 ± 0.14 [a]	44.59 ± 0.62 [a]	0.73 ± 0.01 [a]	73.4 ± 0.06 [a]
Svevo SSIIa	2016	67.01 ± 0.68 [b]	2.13 ± 0.66 [b]	38.31 ± 1.05 [b]	2.06 ± 0.01 [b]	67.3 ± 1.42 [b]
Svevo	2017	71.57 ± 0.17 [c]	−1.95 ± 0.05 [c]	49.06 ± 0.45 [c]	0.21 ± 0.02 [c]	75.6 ± 0.78 [a]
Svevo SBEIIa	2017	64.49 ± 0.92 [d]	1.53 ± 0.27 [d,b]	39.99 ± 1.36 [d]	7.36 ± 0.10 [d]	66.4 ± 0.50 [b]
Commercial wholemeal		44.66	15.25	18.66	nd	nd

DP-L * = dry pasta lightness; DP-a * = dry pasta redness; DP-b * = dry pasta yellowness; RS% = percentage resistant starch; TS = total starch. Numbers in the same column with different superscript letters indicate significant differences (*p* < 0.05).

SBEIIa   Svevo 2017   SSIIa   Wholemeal

Figure 1. Spaghetti appearance of genotypes compared to commercial wholemeal pasta. From left to right SBEIIa, Svevo 2017, SSIIa, Commercial Wholemeal.

Analysis of the colour parameters of the dry pasta show the HA samples are a little duller (lower L*), with more redness (+ve a* values) and are less yellow than their Svevo pasta controls (Table 2) consistent with the visual appearance. There were subtle differences between Svevo in 2016 and 2017 seasons and this is due to environmental effects on colour of pasta or semolina [41]. The SBEIIa semolina was duller and appeared to be browner than Svevo and SSIIa. This could be related to the higher ash content, as noted for HA durum [24]. Indeed, we found higher ash contents in the HA semolina compared with Svevo, consistent with the higher amylose content correlated to lipid amount (Table 1). The duller HA pasta is likely due to their higher protein content since noodle brightness is negatively correlated to protein content [42] but also more bran flecks in the semolina. Clearly the HA pasta have a much more desirable appearance than the commercial wholemeal example we selected from local supermarket, which is much redder, duller, and less yellow. In agreement with our results, Hazard et al. [43] observed a deterioration in pasta color made from high-amylose semolina from SBEII mutants.

Pasta cooking time is affected by the shape, diameter and density of the strand and cooking method, which were controlled in this study. Cooking time is very much influenced by the rate of water migration into the strands and degree of starch gelatinisation, which can be affected by starch composition and starch swelling [39,44]. Both HA pasta had reduced FCT with SSIIa having the lowest (Table 3). This is most likely due the HA pasta having less starch to gelatinise, although this does not explain why SSIIa has shorter FCT than SBEIIa. It is likely the reduced starch swelling (lower swelling power, Table 1) caused by less amylopectin is responsible since it would allow more rapid gelatinisation (lower onset) and shorten cooking time.

Table 3. Pasta cooking properties of HA and control genotypes.

Pasta	Field Season	FCT (s)	Firmness			Overcook Tolerance	Stickiness		Cooking Loss (%)	Water Absorption
			PH (g)	Area (g/s)	PH/P		PH (g)	Area (g/s)		
Svevo	2016	661 ± 50	1334 ± 65 [a]	615 ± 22 [a]	103 ± 5.1 [a]	52 [a]	17.0 ± 1.8 [a,b]	9.3 ± 0.68 [a]	4.6 ± 0.11 [a]	146 ± 2.96 [a]
Svevo SSIIa	2016	578 ± 20	1139 ± 82 [b]	566 ± 51 [b]	80 ± 5.8 [b]	52 [a]	14.7 ± 1.5 [a]	5.4 ± 0.83 [b]	6.9 ± 0.01 [b]	123 ± 0.01 [b]
Svevo	2017	761 ± 19	1261 ± 37 [c]	539 ± 16 [c]	99 ± 2.9 [c]	54 [b]	18.7 ± 1.5 [b]	6.8 ± 2.2 [c,b]	4.8 ± 0.30 [a]	158 ± 0.38 [c]
Svevo SBEIIa	2017	678 ± 40	1124 ± 82 [b]	544 ± 51 [c]	73 ± 5.3 [d]	49 [c]	15.9 ± 1.3 [a]	5.7 ± 1.0 [c,b]	6.6 ± 0.14 [b]	120 ± 1.53 [b]

Data are mean ± stdev. FCT = Fully cooked time; PH = peak height; PH/P = PH divided by semolina protein. Numbers in the same column with different superscript letters indicate significant differences ($p < 0.05$).

Indeed, onset gelatinisation (T_{onset}) of HA pasta was at a lower temperature for SBEIIa but ~6–7 °C lower for SSIIa (Table 4), while end set gelatinisation (T_{end}) was much higher for the HA pasta's especially SBEIIa. Lower gelatinisation enthalpy (ΔH) and greater gelatinisation range for high amylose mutant starches compared to the corresponding wild-type control in material with a wide range in amylose content (36–93%) has been reported [28] confirming our results. We found in the Svevo HA pasta no significant decrease in enthalpy, although it tended to decline but a very wide range in gelatinisation temperature compared to Svevo.

Table 4. Starch gelatinisation properties of genotypes.

	T_{onset} (°C)	T_{peak} (°C)	T_{end} (°C)	Enthalpy (J/kg)
Svevo 2017	52.76 ± 0.46 [b,c]	59.51 ± 0.47 [a,b]	66.78 ± 0.05 [a]	11.71 ± 0.38 [b,c]
Svevo SBEIIa	52.27 ± 0.19 [b]	65.49 ± 0.23 [d]	88.83 ± 0.23 [d]	7.92 ± 1.40 [a,b]
Svevo 2016	53.66 ± 0.56 [c]	59.91 ± 0.19 [b,c]	67.81 ± 0.53 [a,b]	10.91 ± 1.42 [a,b,c]
Svevo SSIIa	47.04 ± 0.08 [a]	60.93 ± 0.40 [c]	69.57 ± 0.18 [c]	7.23 ± 1.50 [a]

Numbers in the same column with different superscript letters indicate significant differences ($p < 0.05$).

The texture analyser compression test tries to evaluate the sensory equivalent of the "first bite" (compression of the strand by the incisors) and the force versus time curve provides two parameters, the height at the peak force and the work to cut or area under the peak (Figure 2). Pasta firmness peak height for the HA lines was reduced compared to Svevo while area under the compression curve was lower in SSIIa compared to its control, but not for the SBEIIa and its control (Table 3). Firmness is correlated to protein content [45,46] and one would expect a higher firmness in the HA pastas. However, adjustment for this shows that peak height/protein (PH/P) for the HA pastas was still significantly lower than the Svevo controls while both HA pasta had similar firmness with SBEIIa significantly softer. While the protein content of the two Svevo semolina samples was almost identical, pasta firmness was lower for Svevo 2017 as other factors come into play, affecting firmness beyond experimental error. The HA pastas have less amylopectin, which would reduce swelling, and this was noted in significantly lower pasta swelling (Table 1) or water absorption (Table 3). Reduced swelling and higher protein content would be expected to make the cooked firmness increase [47] but the opposite occurred. This was in contrast to what has been observed in other works, where an increased firmness was reported in pasta prepared with high amylose semolina [24,37,43,48]. However, Aravind et al. [9] added commercial RSII or RSIII to pasta with no clear effect on pasta firmness. Sozer et al. [44] added green banana with high RS to pasta with no effect on hardness. It is not easy to explain these differences, but sample differences and methods of assessing firmness could be explanations. It is also possible this type of measurement does not reveal the full impacts of high amylose starch in pasta and perhaps a better measure would be the elasticity of the pasta.

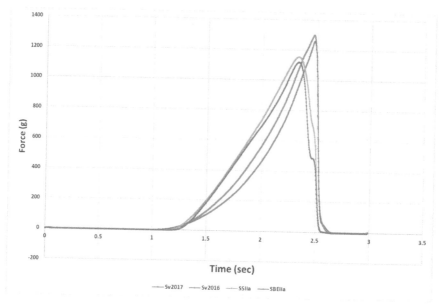

Figure 2. Typical profile for firmness measurement of HA and control genotypes.

Typically consumers overcook pasta and this leads to a reduction in firmness as any ungelatinised starch in the central core becomes fully gelatinised and the pasta can swell more making it softer. The overcooking tolerance or resistance to firmness reduction is a good measure of tolerance and pasta should resist overcooking while still retaining al dente (having some firmness to the bite), and the lower this value, the more tolerant the pasta is to firmness loss due to overcooking, which is desirable. There were small but significant differences between the pasta with SBEIIa having the best tolerance, while the SSIIa pasta overcooking tolerance was not different to its Svevo control (Table 3). Hogg and colleagues [24] noted its high amylose durum (SSIIa null) was found to be more resistant to overcooking compared with the wild type while our Svevo SSIIa pasta had same tolerance to Svevo 2016.

There was a tendency for the HA pastas to have reduced stickiness with SSIIa pasta having lower area, but not significantly different peak height, although tending to be lower, compared to Svevo. While SBEIIa showed the opposite trend with a lower peak height, but not area compared to its Svevo control. Both HA pastas had equivalent stickiness. Soh et al. [37] using reconstitution studies found no change in pasta stickiness made from high amylose maize starch (27–74% amylose), whereas lowering the amylose content (from 23% to 0.7%) makes pasta stickier [49]. Aravind et al. [9] added commercial RSII and RSIII to pasta formulations and found no impact on pasta stickiness. Both HA pasta had significantly higher cooking loss than their Svevo controls with no significant difference between SSIIa and SBEIIa pasta (Table 3). However, the magnitude of the cooking loss is at an acceptable level (7–8%). A higher cooking loss was observed in SBEIIa nulls by Hazard et al. [43], but again they report around 6.4–6.7% results, similar to our own. The higher cooking loss could be related to the higher amylose content and its ability to leach out of the pasta during cooking [43], but this did not lead to increased pasta stickiness. Higher amylose in the pasta also affected water uptake in the pasta being significantly lower than the respective Svevo controls. This could be related to the reduced tendency for HA starch granules to swell as they contain less amylopectin and have tightly packed granules that are more resistant to swelling [50].

Overall, the HA pastas show slightly reduced firmness and increased cooking loss with inferior colour to Svevo but with reduced FCT and slightly lower stickiness. In conclusion, both highlight an acceptable quality. The colour may be an issue and require further breeding to improve but these

HA pastas are superior to the commercial wholemeal pasta chosen for comparison. Nevertheless, the HA pasta improves RS and hence dietary fibre by 128% without the need for adding bran and would achieve more consumer acceptance being similar in appearance to regular, normal amylose content pasta. The nutritional value of pasta from both HA lines are improved in terms of their increase in resistant starch, which is a fibre, and lowering of the GI (for SBEIIa), see below. The total dietary fibre content of Svevo 2017 and SBEIIa were measured and showed significant increase with SBEIIa consisting of 67% insoluble and 33% soluble fibre (Table 1). In every 100 g dried pasta cooked, there is 5.7 g total dietary fibre which meets 23–19% of the recommended fibre daily requirement (2–30 g/d) (https://www.nutritionaustralia.org/national/resource/fibre). Further work is required using a trained sensory panel to determine the differences between the pastas and a consumer panel to determine the preferences for the four different pastas (standard, SSIIa, SBEIIa, wholemeal).

3.2. Impact of Elevated Amylose on Pasta In Vitro Starch Digestion

The HA pasta's had elevated RS compared to Svevo, with SBEIIa achieving a 35 fold increase relative to Svevo 2017 while SSIIa achieved a 2.8-fold increase relative to Svevo 2016 (Table 2). Elevation of RS content in endosperm can be achieved by reducing SSIIa activity, enhanced GBSSI activity or hindered SBEII activity [51]. We were interested to know the starch digestion kinetics of the HA pasta samples that allows to quantify rate and extent of digestion which can be achieved using an in vitro method. Based on other studies [52,53], we expect a reduction in the extent of starch digestion in the HA pastas compared to Svevo. Typical starch digestion curves over the 360 min digestion are shown in Figure 3. Initially, digestion is rapid with little difference between the pastas up to ~30 min, followed by a lower rate with the data following an exponential curve. Up to 50 min into the starch digestion, only SBEIIa was showing signs of slowing its rate and this continued reaching a lower final digestion compared to Svevo. The curves for the two Svevo pasta's overlap, as expected while SSIIa and especially the SBEIIa, show lower rates and extent of digestion. For example, after 250 min <40% of starch was digested for SBEIIa while ~70% is digested in Svevo and ~58% in SSIIa. These plots are typical for pasta and the slow starch digestion in pasta is due to the compact microstructure of pasta and because the starch is embedded in a gluten matrix [32,54].

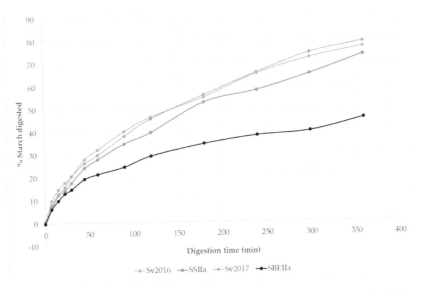

Figure 3. Digestibility curves obtained for cooked pasta of HA and control genotypes digested by α amylase/amyloglucosidase.

The transformation of the data into logarithm of slope plots reveals two distinct linear steps where ~20–40% of starch is digested in the first phase which is faster (k_1) and the remainder of the starch digested in a second stage, with a slower rate constant (k_2) (Table 5). The areas under the curves (AUC) allow a comparison of the extent of starch digestion in the pasta samples and slower digestion would appear as lower AUC values. The AUC values reflect these changes and can be normalised to the AUC in Svevo 2016. For the two Svevo samples, the k_1 and k_2 values are very similar as are the AUC and shows that despite different growing season, this did not impact on starch digestion significantly. However, both HA samples have significantly lower AUC than Svevo controls with SBEIIa having the lowest AUC by a significant margin (Table 4). Interestingly, k_1 and k_2 values for SBEIIa are faster than all samples but the extent of digestion ($C\infty$ %) is much reduced, suggesting that what is available for digestion is more rapidly digested and that not digested is resistant starch. To the best of our knowledge, there are no in vitro studies focused on the digestion of high amylose pasta.

Table 5. Kinetic parameters of digestibility of different pasta products.

Pasta	Total Area under Digestion Curve	Normalised Area	$C\infty$ % I	k_1	$C\infty$ % II	k_2
Svevo 2016	19,851 [a]	1.00	38.9	0.02756	83.2	0.00654
Svevo SSIIa	18,652 [b]	0.94	35.4	0.02554	81.0	0.00574
Svevo 2017	20,572 [a]	1.04 (1.00)	37.2	0.02670	87.1	0.00615
Svevo SBEIIa	12231 [c]	0.62 (0.59)	24.3	0.03470	45.2	0.00831

Numbers in the same column with different superscript letters indicate significant differences ($p < 0.05$). k_1 and k_2 refer to starch digestion rate constants at each phase; $C\infty$ % I and $C\infty$ % II refer to estimated % of starch digested at each phase.

Hoebler et al. [55] showed that bread rich in amylose had a lower starch degradation compared to control bread. Corrado et al. [56] reported that the starch amylolysis rate and extent were lower for SBEIIa/b-AB compared to those of the control. In rice, the starch digestion rate (k) decreased in flour blend with an increased amount of the HA maize starch [57]. The reason why starch digestibility is reduced with higher amylose content could be due to the strong interaction among the linear polymers of amylose (retrogradation) and between amylose and lipids that results in complex formation on the surface of starch granules [58].

3.3. Impact of Elevated Amylose on Pasta In Vivo Starch Digestion and Glycaemic Index

The best measure of a foods glycaemic index is to perform a test with human subjects. The GI test was developed to rank equal carbohydrate portions of different foods according to the extent to which they increase blood glucose levels after the test food is ingested and glucose moves from the intestines into the cardiovascular system. High GI foods are rapidly digested and this produces a sudden and large spike in blood glucose followed by a sharp fall, often achieving a lower blood glucose than basal (pre-ingestion) glucose. In contrast, a low GI food is more slowly digested and this results in a more gradual and lower elevation of blood glucose. Dieticians are using food GI values in planning diets suitable for people with diabetes. Long term epidemiological studies have shown that consumption of high GI impact foods which causes surges in blood glucose and insulin levels, increases the risk of developing type II diabetes, heart disease, obesity, and certain cancers [59,60]. Therefore, the development of low-GI foods could assist with the prevention and treatment of these diseases. In our study, we prepared the semolina and pasta under the same processing conditions, with only water and semolina in the pasta mix, since these factors can impact a foods GI, so only compositional differences matter, which in this case is the amount of amylose as the glutenin composition is identical amongst the genotypes having a glutenin Glu-A1 null, Glu-B1 7 + 8, and LMW-2 (data not shown).

Test meal characteristics are reported in Table 6 with a larger portion size of the HA pasta needed to achieve equivalent available starch content. Our in vivo GI testing results are depicted in Figure 4 and Table 7. The reference food (glucose) produced a rapid rise in plasma glucose to a high peak concentration at 30 min then declined reaching below pre-fasting levels after 110–118 min. For the pastas, the glycaemic responses showed a steady rise and much lower peak maxima accompanied by a more gentle return to almost basal glucose levels by 120 min. Svevo 2016 and SSIIa pasta showed similar behaviour in their plasma glucose responses with SSIIa showing slightly lower glucose values between 50–120 min (Figure 4A). Comparing Svevo 2017 with SBEIIa pasta, again we observed similar responses in plasma glucose, but this time, the rise in the SBEIIa pasta was lower and glucose was at a lower level than Svevo 2017 (control) from 15 to 120 min (Figure 4B). Calculation of the mean (10 subjects) GI values for the samples is shown in Table 7. The reference food's GI value (100) was significantly greater than the average GI for both pasta products ($p < 0.001$). Comparing Svevo 2016 with SSIIa, there was no significant difference in GI ($p < 0.05$) despite showing a lower GI for the latter and being significantly lower in the in vitro assay (Table 5). However, comparing Svevo 2017 with SBEIIa, there was a significantly lower GI in the latter by 10 units. There was no difference in GI between the two Svevo pasta samples, showing that seasonal conditions had no real impact on pasta.

GI ($p = 0.015$) as was noted for the in vitro data. This is the first report of pasta made from 100% durum wheat with a reduction in the GI due to genetic difference in the amylose content. Previous studies have indicated that elevated amylose contents are associated with higher levels of resistant starch in wheat based products (bread and rusks) that induce lower postprandial glycaemic responses [61]. Foods with a GI < 55 are considered to be low-GI foods, which includes all the four pasta samples but the SBEIIa has an even lower GI which is a good achievement.

Table 6. The weight and carbohydrate contents of the test portions of the reference food (glucose) and the HA and control pasta products, calculated using manufacture's data.

Test Food	Available Carbohydrates for 100 Grams (g)	Portion Size (g)	Available Carbohydrates in Test Portion (g)
Glucose (ref.)	97.30	51.4 g glucose 250 mL water	50.0
Svevo 2016	64.19	77.9 g dry pasta	50.0
Svevo SSIIa	57.35	87.2 g dry pasta	50.0
Svevo 2017	74.05	67.5 g dry pasta	50.0
Svevo SBEIIa	65.27	76.6 g dry pasta	50.0

Table 7. Glycaemic index (GI) and glycaemic load (GL) of the reference food and the two pasta samples.

Pasta Sample (Amylose%)	GI	GL
Svevo 2016 (33)	52 ± 3 [a]	17
Svevo SSIIa (44)	49 ± 3 [a]	14
Glucose	100 ± 0 [b]	
Svevo 2017 (32)	48 ± 4 [a]	18
Svevo SBEIIa (58)	38 ± 3 [c]	12

Data are mean ± SE. Values with alike superscript letters in the same column are not different, $p < 0.05$.

(A)

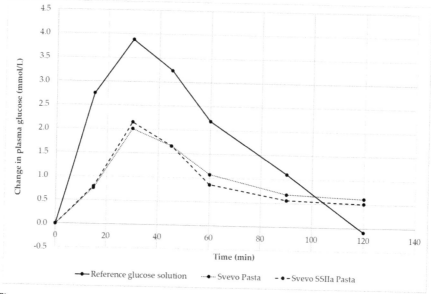

(B)

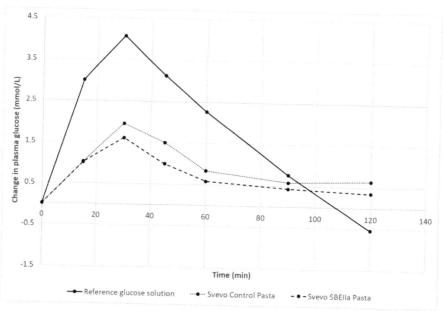

Figure 4. The average plasma glucose response curves (n = 10 subjects) for the equal-carbohydrate portions of the reference food and the two pasta products. (**A**): Svevo 2016 and Svevo SSIIa; (**B**): Svevo 2016 and Svevo SBEIIa, shown as the change in plasma glucose from the fasting baseline level.

There were significant correlations between amylose content, resistant starch, in vitro AUC with each other and importantly, with in vivo GI (r 0.90–0.93, $p < 0.05$) but sample size is understandably low due to the high cost of GI testing. Nevertheless, we feel these correlations would prevail even if sample size increased markedly. The higher the amylose, the more resistant starch and the lower the in vitro AUC and extent of digestion. The lower the AUC determined in the in vitro assay the lower the GI with a good prediction (r 0.93). This correspondence between the in vitro and in vivo methods has been shown elsewhere [62]. For more practical purposes such as food consumption, if the amount of these pastas consumed exceeds 50 g, it is better to determine the glycaemic load (GL) value for any sized carbohydrate-containing food. GL = [(GI × amount of available carbohydrate in the portion (50 g) of pasta)/100] and the lower the better, these are shown in Table 7. Dietary intervention studies show that the digestibility of high amylose starch is generally lower than normal starch [28]. As such, foods containing high amylose starch, like the two pastas described here, should help reduce the risk of developing type 2 diabetes. The development and commercialisation of high amylose wheat-based products will allow new products with higher fibre and better sensory acceptability than current wholemeal and wholegrain wheat products designed to increase fibre. Recently, Corrado [56] reported a semolina pudding made from a SBEIIa durum wheat mutant with resistant starch content of ~5%, digested more slowly and to a lesser extent by in vitro test than wild type control. However, they found no difference in the foods GI. Pasta is a low GI traditional Italian food with a crucial role in Mediterranean diet. Pasta consumption has been associated to beneficial effects for human health compared with higher-GI dietary foods. Recently, in a study [63], it has been reported that pasta consumption, in the context of low glycemic index dietary patterns, has no negative effect on body weight and mass. Here, the in vitro and in vivo studies demonstrated that the nutritional value of the pasta was significantly improved with a further reduction of the glycemic index in one of the HA pastas. As the consumption of pasta is constantly growing not only in developed but developing countries, the HA durum wheat genotypes open interesting perspectives in the use of HA pasta products as a vehicle for the prevention of serious non-transmissible diseases, such as colon cancer, type 2 diabetes, and cardiovascular disorders.

4. Conclusions

This is the first study of the glycaemic response of pasta prepared from SBEIIa and SSIIa durum wheat mutants with elevated amylose and resistant starch content in a processed pasta food matrix. While the SSIIa mutant increased amylose content, it did not significantly lower spaghetti GI while the SBEIIa mutant, with higher amylose and resistant starch content in the spaghetti, did lower GI. This suggests that a minimum amylose or resistant starch content is needed before GI is lowered in the spaghetti food matrix. We suggest a minimum RS content in spaghetti of ~7% which corresponded to an amylose content of ~58% is needed to lower GI. Both the HA pastas provide higher dietary fibre and resistant starch compared to Svevo and have minimal impacts on pasta technological properties. While the colour of the HA pastas were inferior to Svevo, they were much closer to 100% durum semolina pasta than a typical commercial wholemeal pasta. However, detailed sensory analysis is needed to determine consumer acceptability of this spaghetti. Further evaluation across more environments could be useful. It might then be necessary to further develop the mutants by crossing to improve some of the minor quality deficiencies, but they would have to be re-evaluated for GI and other benefits in future studies, e.g., satiety and long term glyacemic benefits. Compared to typical commercial wholemeal pasta, these HA pastas would be much more visually appealing while providing additional nutritional benefits.

Supplementary Materials: The following are available online at http://www.mdpi.com/2304-8158/9/6/693/s1, Table S1: Decadal values of temperature and precipitation recorded in the experimental field from November 2015/2016 to June 2016/2017.

Author Contributions: Conceptualization, M.S. and D.L.; Formal analysis, M.S. and F.S.; Funding acquisition, M.S. and D.L.; Investigation, M.S. and D.L.; Methodology, M.S., E.B., S.M., and D.L.; Project administration,

M.S. and D.L.; Resources, M.S. and D.L.; Supervision, M.S. and D.L.; Writing—original draft, M.S.; Writing—review & editing, F.S., E.B., S.M., and D.L. All authors have read and agreed to the published version of the manuscript.

Funding: This research was funded by AGER "Agroalimentare e Ricerca" (From Seed to Pasta project) and by the Italian Ministry of Education, University, and Research (MIUR) in the frame of the MIUR initiative "Departments of excellence", Law 232/2016 and supported by the New South Wales, Department of Primary Industries.

Acknowledgments: The authors would like to thank Fiona Atkinson of the Sydney University Glycemic Index Research Service for conducting the GI testing studies and Alessandro Cammerata of Council for Agricultural Research and Analysis of Agricultural Economics-Engineering and Agro-Food Processing (CREA-IT) for semolina production. We would also like to than Denise Pleming (NSW DPI) for MicroDoughLab measurements and Caili Li (University of Queensland) for the DSC measurements.

Conflicts of Interest: The authors declare no conflict of interest. The funders had no role in the design of the study, in the collection, analyses, or interpretation of data, in the writing of the manuscript, or in the decision to publish the results.

References

1. Prückler, M.; Siebenhandl-Ehn, S.; Apprich, S.; Höltinger, S.; Haas, C.; Schmid, E.; Kneifel, W. Wheat bran-based biorefinery 1: Composition of wheat bran and strategies of functionalization. *LWT* **2014**, *56*, 211–221. [CrossRef]

2. Björck, I.; Östman, E.; Kristensen, M.; Anson, N.M.; Price, R.K.; Haenen, G.R.; Havenaar, R.; Knudsen, K.E.B.; Frid, A.; Mykkänenh, H.; et al. Cereal grains for nutrition and health benefits: Overview of results from in vitro, animal and human studies in the HEALTHGRAIN project. *Trends Food Sci. Tech.* **2012**, *25*, 87–100. [CrossRef]

3. Zong, G.; Gao, A.; Hu, F.B.; Sun, Q. Whole grain intake and mortality from all causes, cardiovascular disease, and cancer a meta-analysis of prospective cohort studies. *Circulation* **2016**, *133*, 2370–2380. [CrossRef] [PubMed]

4. Jones, J.M. CODEX-aligned dietary fiber definitions help to bridge the 'fiber gap'. *Nut. J.* **2014**, *13*, 34. [CrossRef] [PubMed]

5. Grigor, J.M.; Brennan, C.S.; Hutchings, S.C.; Rowlands, D.S. The sensory acceptance of fibre-enriched cereal foods: A meta-analysis. *Int. J. Food Sci. Technol.* **2016**, *51*, 3–13. [CrossRef]

6. Mercier, S.; Moresoli, C.; Mondor, M.; Villeneuve, S.; Marcos, B. A meta-analysis of enriched pasta: What are the effects of enrichment and process specifications on the quality attributes of pasta? *Compr. Rev. Food Sci. Food Saf.* **2016**, *15*, 685–704. [CrossRef]

7. Bird, A.R.; Regina, A. High amylose wheat: A platform for delivering human health benefits. *J. Cereal Sci.* **2018**, *82*, 99–105. [CrossRef]

8. Sajilata, M.G.; Singhal, R.S.; Kulkarni, P.R. Resistant starch—A review. *Compr. Rev. Food Sci. Food Saf.* **2006**, *5*, 1–17. [CrossRef]

9. Aravind, N.; Sissons, M.; Fellows, C.M.; Blazek, J.; Gilbert, E.P. Optimisation of resistant starch II and III levels in durum wheat pasta to reduce in vitro digestibility while maintaining processing and sensory characteristics. *Food Chem.* **2013**, *136*, 1100–1109. [CrossRef]

10. Bustos, M.C.; Perez, G.T.; Leon, A.E. Sensory and nutritional attributes of fibre-enriched pasta. *LWT* **2011**, *44*, 1429–1434. [CrossRef]

11. Gelencsér, T.; Gál, V.; Hódsagi, M.; Salgó, A. Evaluation of quality and digestibility characteristics of resistant starch-enriched pasta. *Food Bioprocess Tec.* **2008**, *1*, 171–179. [CrossRef]

12. Sozer, N.; Dalgic, A.C.; Kaya, A. Thermal, textural and cooking properties of spaghetti enriched with resistant starch. *J. Food Eng.* **2007**, *81*, 476–484. [CrossRef]

13. Berry, C.S. Resistant starch. Formation and measurement of starch that survives exhaustive digestion with amylolytic enzymes during the determination of dietary fiber. *J. Cereal Sci.* **1986**, *4*, 301–314. [CrossRef]

14. Li, H.; Gidley, M.J.; Dhital, S. High-amylose starches to bridge the "fiber gap": Development, structure, and nutritional functionality. *Compr. Rev. Food Sci. Food Saf.* **2019**, *18*, 362–379. [CrossRef]

15. Regina, A.; Bird, A.; Topping, D.; Bowden, S.; Freeman, J.; Barsby, T.; Kosar-Hashemi, B.; Li, Z.; Rahman, S.; Morell, M. High-amylose wheat generated by RNA interference improves indices of large-bowel health in rats. *Proc. Natl. Acad. Sci. USA* **2006**, *103*, 3546–3551. [CrossRef] [PubMed]

16. Slade, A.J.; Fuerstenberg, S.I.; Loeffler, D.; Steine, M.N.; Facciotti, D. A reverse genetic, nontransgenic approach to wheat crop improvement by TILLING. *Nat. Biotechnol.* **2005**, *23*, 75–81. [CrossRef] [PubMed]
17. Slade, A.J.; McGuire, C.; Loeffler, D.; Mullenberg, J.; Skinner, W.; Fazio, G.; Knauf, V.C. Development of high amylose wheat through TILLING. *BMC Plant Biol.* **2012**, *12*, 69. [CrossRef]
18. Sestili, F.; Janni, M.; Doherty, A.; Botticella, E.; D'Ovidio, R.; Masci, S.; Jones, H.D.; Lafiandra, D. Increasing the amylose content of durum wheat through silencing of the SBEIIa genes. *BMC Plant Biol.* **2010**, *10*, 144. [CrossRef]
19. Sestili, F.; Palombieri, S.; Botticella, E.; Mantovani, P.; Bovina, R.; Lafiandra, D. TILLING mutants of durum wheat result in a high amylose phenotype and provide information on alternative splicing mechanisms. *Plant Sci.* **2015**, *233*, 127–133. [CrossRef]
20. Botticella, E.; Sestili, F.; Sparla, F.; Moscatello, S.; Marri, L.; Cuesta-Seijo, J.A.; Falini, G.; Battistelli, A.; Trost, P.; Lafiandra, D. Combining mutations at genes encoding key enzymes involved in starch synthesis affects the amylose content, carbohydrate allocation and hardness in the wheat grain. *Plant Biotechnol. J.* **2018**, *16*, 1723–1734. [CrossRef]
21. Hazard, B.; Zhang, X.; Colasuonno, P.; Uauy, C.; Beckles, D.M.; Dubcovsky, J. Induced mutations in the starch branching enzyme II (SBEII) genes increase amylose and resistant starch content in durum wheat. *Crop Sci.* **2012**, *52*, 1754–1766. [PubMed]
22. Botticella, E.; Sestili, F.; Ferrazzano, G.; Mantovani, P.; Cammerata, A.; D'Egidio, M.G.; Lafiandra, D. The impact of the SSIIa null mutations on grain traits and composition in durum wheat. *Breed. Sci.* **2016**, *66*, 572–579. [CrossRef] [PubMed]
23. Hogg, A.C.; Gause, K.; Hofer, P.; Martin, J.M.; Graybosch, R.A.; Hansen, L.E.; Giroux, M.J. Creation of a high-amylose durum wheat through mutagenesis of starch synthase II (SSIIa). *J. Cereal Sci.* **2013**, *57*, 377–383. [CrossRef]
24. Hogg, A.C.; Martin, J.M.; Manthey, F.A.; Giroux, M.J. Nutritional and quality traits of pasta made from SSIIa null high-amylose durum wheat. *Cereal Chem.* **2015**, *92*, 395–400. [CrossRef]
25. AACC International. *Approved Methods of Analysis*, 11th ed.; Method 26-41.01., 44-15A., [25]54-70.01., 66-51.01; AACC International: St. Paul, MN, USA, 2010.
26. Horneck, D.A.; Miller, R.O. Determination of total nitrogen in plant tissue. *Handb. Ref. Methods Plant Anal.* **1998**, *2*, 75–83.
27. Sharma, R.; Sissons, M.J.; Rathjen, A.J.; Jenner, C.F. The null-4A allele at the *waxy* locus in durum wheat affects pasta cooking quality. *J. Cereal Sci.* **2002**, *35*, 287–297. [CrossRef]
28. Li, H.; Dhital, S.; Slade, A.J.; Yu, W.; Gilbert, R.G.; Gidley, M.J. Altering starch branching enzymes in wheat generates high-amylose starch with novel molecular structure and functional properties. *Food Hydrocoll.* **2019**, *92*, 51–59. [CrossRef]
29. Sissons, M.; Ovenden, B.; Adorada, D.; Milgate, A. Durum wheat quality in high input irrigation systems in south eastern Australia. *Crop Pasture Sci.* **2014**, *65*, 411–422. [CrossRef]
30. Morita, N.; Maeda, T.; Miyazaki, M.; Yamammori, M.; Miura, H.; Phtsuka, I. Dough and baking properties of high amylose and waxy wheat flours. *Cereal Chem.* **2002**, *79*, 491–495. [CrossRef]
31. Sissons, M.J.; Aravind, N.; Fellows, C.M. Quality of fibre-enriched spaghetti containing microbial transglutaminase. *Cereal Chem.* **2010**, *87*, 57–64. [CrossRef]
32. Zou, W.; Sissons, M.; Gidley, M.J.; Gilbert, R.G.; Warren, F.J. Combined techniques for characterising pasta structure reveals how the gluten network slows enzymic digestion rate. *Food Chem.* **2015**, *188*, 559–568. [CrossRef] [PubMed]
33. Vansteelandt, J.; Delcour, J.A. Characterisation of starch from durum wheat (Triticum durum). *Starch-Stärke* **1999**, *51*, 73–80. [CrossRef]
34. Hogg, A.C.; Martin, J.M.; Giroux, M.J. Novel *ssIIa* alleles produce specific seed amylose levels in hexaploid wheat. *Cereal Chem.* **2017**, *94*, 1008–1015. [CrossRef]
35. Rakszegi, M.; Kisgyörgy, B.N.; Kiss, T.; Sestili, F.; Láng, L.; Lafiandra, D.; Bedő, Z. Development and characterization of high-amylose wheat lines. *Starch-Stärke* **2015**, *67*, 247–254. [CrossRef]
36. Konik-Rose, C.; Thistleton, J.; Chanvrier, H.; Tan, I.; Halley, P.; Gidley, M.; Kosar-Hashemi, B.; Wang, H.; Larroque, O.; Ikea, J.; et al. Effects of starch synthase IIa gene dosage on grain, protein and starch in endosperm of wheat. *Theor. Appl. Genet.* **2007**, *115*, 1053–1062. [CrossRef]

37. Soh, H.N.; Sissons, M.J.; Turner, M.A. Effect of starch granule size distribution and elevated amylose content on durum dough rheology and spaghetti cooking quality. *Cereal Chem.* **2006**, *83*, 513–519. [CrossRef]

38. Tester, R.F.; Morrison, W.R. Swelling and gelatinization of cereal starches. I. Effects of amylopectin, amylose, and lipids. *Cereal Chem.* **1990**, *67*, 551–557.

39. Tomoko, S.; Matsuki, J. Effect of wheat starch structure on swelling power. *Cereal Chem.* **1998**, *75*, 525–529.

40. Dick, J.W.; Youngs, V.L. Evaluation of durum wheat, semolina, and pasta in the United States. In *Durum Wheat: Chemistry and Technology*; Fabriani, G., Lintas, C., Eds.; American Association of Cereal Chemists: St. Paul, MN, USA, 1988; pp. 237–248.

41. Clarke, F.R.; Clarke, J.M.; McCaig, T.N.; Know, R.E.; DePauw, R.M. Inheritance of yellow pigment concentration in seven durum wheat crosses. *Can. J. Plant Sci.* **2006**, *86*, 133–141. [CrossRef]

42. Wang, C.; Kovacs, M.I.P.; Fowler, D.B.; Holley, R. Effects of protein content and composition on white noodle making quality: Color. *Cereal Chem.* **2004**, *81*, 777–784. [CrossRef]

43. Hazard, B.; Zhang, X.; Naemeh, R.; Hamilton, M.K.; Rust, B.; Raybould, H.E.; Newman, J.W.; Martin, R.; Dubcovsky, J. Mutations in durum wheat SBEII genes conferring increased amylose and resistant starch affect grain yield components, semolina and pasta quality and fermentation responses in rats. *Crop Sci.* **2015**, *55*, 2813–2825. [CrossRef] [PubMed]

44. Sozer, N.; Kaya, A. The effect of cooking water composition on textural and cooking properties of spaghetti. *Int. J. Food Prop.* **2008**, *11*, 351–362. [CrossRef]

45. Edwards, N.M.; Izydorczyk, M.S.; Dexter, J.E.; Biliaderis, C.G. Cooked pasta texture: Comparison of dynamic viscoelastic properties to instrumental assessment of firmness. *Cereal Chem.* **1993**, *70*, 122–126.

46. Sissons, M.J.; Egan, N.E.; Gianibelli, M.C. New insights into the role of gluten on durum pasta quality using reconstitution method. *Cereal Chem.* **2005**, *82*, 601–608. [CrossRef]

47. Martin, J.M.; Hogg, A.C.; Hofer, P.; Manthey, F.A.; Giroux, M.J. Impacts of *SSIIa-A* null allele on durum wheat noodle quality. *Cereal Chem.* **2014**, *91*, 176–182. [CrossRef]

48. Martin, J.M.; Talbert, L.E.; Habernicht, D.K.; Lanning, S.P.; Sherman, J.D.; Carlson, G.; Giroux, M.J. Reduced amylose effects on bread and white salted noodle quality. *Cereal Chem.* **2004**, *81*, 188–193. [CrossRef]

49. Gianibelli, M.C.; Sissons, M.J.; Batey, I.L. Effect of different waxy starches on pasta cooking quality of durum wheat. *Cereal Chem.* **2005**, *82*, 321–327. [CrossRef]

50. Haralampu, S.G. Resistant starch—A review of the physical properties and biological impact of RS3. *Carbohydr. Polym.* **2000**, *41*, 285–292. [CrossRef]

51. Newberry, M.; Berbezy, P.; Belobrajdic, D.; Chapron, S.; Tabouillot, P.; Regina, A.; Bird, A. High-amylose wheat foods: A new opportunity to meet dietary fiber targets for health. *Cereal Foods World* **2018**, *63*, 188–193.

52. Evans, A.; Thompson, D.B. Resistance to α-amylase digestion in four native high-amylose maize starches. *Cereal Chem.* **2004**, *81*, 31–37. [CrossRef]

53. Tester, R.F.; Qi, X.; Karkalas, J. Hydrolysis of native starches with amylases. *Anim. Feed Sci. Tech.* **2006**, *130*, 39–54. [CrossRef]

54. Fardet, A.; Hoebler, C.; Baldwin, P.M.; Bouchet, B.; Gallant, D.J.; Barry, J.L. Involvement of the protein network in the in vitro degradation of starch from spaghetti and lasagne: A microscopic and enzymic study. *J. Cereal Sci.* **1998**, *27*, 133–145. [CrossRef]

55. Hoebler, C.; Karinthi, A.; Chiron, H.; Champ, M.; Barry, J.L. Bioavailability of starch in bread rich in amylose: Metabolic responses in healthy subjects and starch structure. *Eur. J. Clin. Nutr.* **1999**, *53*, 360–366. [CrossRef] [PubMed]

56. Corrado, M.; Cherta-Murillo, A.; Chambers, E.S.; Wood, A.J.; Plummer, A.; Lovegrove, A.; Edwards, C.H.; Frost, G.S.; Hazard, B.A. Effect of semolina pudding prepared from starch branching enzyme IIa and b mutant wheat on glycaemic response in vitro and in vivo: A randomised controlled pilot study. *Food Funct.* **2020**, *11*, 617–627. [CrossRef] [PubMed]

57. Mir, J.A.; Srikaeo, K.; García, J. Effects of amylose and resistant starch on starch digestibility of rice flours and starches. *Int. Food Res. J.* **2013**, *20*, 1329–1335.

58. Svihus, B.; Uhlen, A.K.; Harstad, O.M. Effect of starch granule structure, associated components and processing on nutritive value of cereal starch: A review. *Anim. Feed Sci. Tech.* **2005**, *122*, 303–320. [CrossRef]

59. Joint FAO/WHO Report. *Carbohydrates in Human Nutrition*; FAO Food and Nutrition, 1998; p. 66. Available online: https://www.who.int/nutrition/publications/nutrientrequirements/9251041148/en/ (accessed on 27 May 2020).

60. Vetrani, C.; Sestili, F.; Vitale, M.; Botticella, E.; Giacco, R.; Griffo, E.; Costabile, G.; Cipriano, P.; Tura, A.; Pacini, G.; et al. Metabolic response to amylose-rich wheat-based rusks in overweight individuals. *Eur. J. Clin. Nutr.* **2018**, *72*, 904–912. [CrossRef]

61. Belobrajdic, D.P.; Regina, A.; Klingner, B.; Zajac, I.; Chapron, S.; Berbezy, P.; Bird, A.R. High-amylose wheat lowers the postprandial glycemic response to bread in healthy adults: A randomized controlled crossover trial. *J. Nutr.* **2019**, *149*, 1335–1345. [CrossRef]

62. Edwards, C.H.; Cochetel, N.; Setterfield, L.; Perez-Moral, N.; Warren, F.J. A single-enzyme system for starch digestibility screening and its relevance to understanding and predicting the glycaemic index of food products. *Food Funct.* **2019**, *10*, 4751–4760. [CrossRef]

63. Chiavaroli, L.; Kendall, C.W.C.; Braunstein, C.R.; Blanco Mejia, S.; Leiter, L.A.; Jenkins, D.J.A.; Sievenpiper, J.L. Effect of pasta in the context of low-glycaemic index dietary patterns on body weight and markers of adiposity: A systematic review and meta-analysis of randomised controlled trials in adults. *BMJ Open* **2018**, *8*, e019438. [CrossRef]

Article

Is Site-Specific Pasta a Prospective Asset for a Short Supply Chain?

Gabriella Pasini [1], Giovanna Visioli [2],* and Francesco Morari [1]

1   Department of Agronomy, Food, Natural Resources, Animals and the Environment, University of Padova,
    Viale dell' Università 16, 35020 Legnaro-Padua, Italy; gabriella.pasini@unipd.it (G.P.);
    francesco.morari@unipd.it (F.M.)
2   Department of Chemistry, Life Sciences and Environmental Sustainability, University of Parma, Parco Area
    delle Scienze 11/a, 43124 Parma, Italy
*   Correspondence: giovanna.visioli@unipr.it; Tel.: +39-0521-905692

Received: 13 March 2020; Accepted: 8 April 2020; Published: 10 April 2020

Abstract: In the 2011–2012 season, variable-rate nitrogen (N) fertilization was applied two times during durum wheat vegetative growth in three field areas which differed in soil fertility in northern Italy. The quality traits of the mono-varietal pasta obtained from each management zone were assessed in view of site-specific pasta production for a short supply chain. To this purpose, semolina from cv. *Biensur* obtained from management zones with different fertility treated with N at variable rate was tested in comparison with a commercial reference (cv. *Aureo*) to produce short-cut pasta. *Biensur* semolina demonstrated to have technological characteristics positively correlated with the low-fertility zones treated with high N doses (200 and 200+15 kg/ha) and, to a lesser extent, with the high-soil-fertility zones (130 and 130 + 15 kg/ha of N). The lower quality parameters were obtained for pasta produced with wheat from medium-fertility zones, independently of the N dose applied. The derived pasta obtained from the low-fertility zones treated with high N doses had cooking and sensory properties comparable to those of pasta obtained using the reference cv. *Aureo*. These results are explained by the higher amounts of gluten proteins and by a higher glutenin/gliadin ratio in semolina, which are indicators of technological quality. Overall, the results indicate that segregation of the grain at harvest led to the production of semolina with higher protein content and, hence, to a higher pasta quality. Therefore, site-specific pasta could be a potential asset for a short supply chain, aiming to improve traceability and environmental and economic sustainability.

Keywords: durum wheat; precision harvest; pasta quality; pasta short chain

1. Introduction

The key determinants of durum wheat flour quality are the quantity and the quality of gluten proteins. These traits are genetically determined but are also influenced by the climatic conditions and the fertility of a cultivated soil. Northern Italy represents the limit for cultivation of durum wheat in Europe, since in this environment, high-quality standards (i.e., protein content) can be achieved by increasing nitrogen (N) inputs, albeit posing environmental risks [1,2]. In particular, in northern Italy, high rainfall associated with shallow water table conditions and alkaline soils increases the risk of N pollution in water and air [3]. In addition, in uniformly managed fields, high yielding areas or areas with low plant-available soil N may result in low grain protein content due to smaller amounts of available N per kg of grain yield [4].

Recent advances in precision agriculture offer new potentialities to meet grain quality standards. Econometric analyses have shown that the gross income of wheat can be maximized by the combination of variable-rate N fertilization (VR-N) with specific quality criteria [5]. VR-N can in fact reduce the risk of low yields by increasing the probability of maintaining the quality of wheat to the appropriate standard [6].

The amount of protein in grains is an essential parameter for obtaining excellent economic results [7], especially when there are contracts that recognize a quality award. The possibility of carrying out a precision harvest of wheat based on its different quality (i.e., protein content) in the field can be a useful system to achieve the protein requirements established by these contracts [8]. In the two-year experiment on VR-N carried out in a previous study [2], only high N input sandy areas met the standard necessary to be eligible for the premium-quality grain protein content (i.e., 13.5%). Conversely, the average protein content of the entire field was always lower the premium threshold.

The variability of protein content within a field is often huge, justifying the possibility of segregating grains of greater or lesser quality. This procedure can be done at post-harvest in the farm using specific sensors [9–11]. Regarding the pre-harvest segregation procedure, it requires the collection of previous information on the quality and yield of cereals [12,13], which could be inferred on the basis of data on management areas for farmers who apply VR-N [14], such as those collected by proximal sensing sensors (e.g., Normalized Difference Vegetation Index).

Another approach for precision harvesting is to use on-the-go sensors, such as Near-Infrared Reflectance mounted on combine harvesters, which allows the analysis of the percentage of protein during grain flow [9,15]. These sensors could potentially allow to automatically separate wheat grains based on their protein (% N) content [16] and thus are encouraged to be adopted.

Though different authors have demonstrated that high profits could be achieved by separating different grain qualities during harvest [17], a study evaluating the effect at field scale on pasta quality of spatial variability and the feasibility to produce "site-specific" pasta for a short supply chain is still lacking. The environmental impact of the entire pasta production cycle, from field to packaging, has been reviewed [18]. The authors highlighted that the production of grain (i.e., wheat cultivation) and semolina were the sub-processes that mostly impacted on the environment. Moreover, under a recent Italian decree [19], the origin of the durum wheat grains used in pasta production must be declared. In light of this, an advantageous strategy would be to promote local, short food supply chains in order to improve environmental and economic sustainability [20]. The idea of implementing a short supply chain by adopting selective harvesting strategies was first put in place in the Australian wine market [21]. These authors were able to identify two harvesting zones in a Cabernet Sauvignon vineyard by the use of remote sensing technologies [21]. By segregating grapes in two bins, they produced wines of different quality and, correspondingly, prices. However, a study [22] questioned the scale at which to apply precision harvesting in order to achieve commercial-scale vinification. Bramley and collaborators [23] confirmed the benefits provided by precision harvesting even when grape and wine production is oriented toward large vinification volumes.

In a previous work, we performed an experiment of VR-N on durum wheat in northern Italy with the aim to obtain high yields and protein contents, in relation to field soil properties and N fertilization. According to our previous results, we advanced the hypothesis that protein variability in grain from different field zones could justify the manufacturing of site-specific pasta as a potential asset for a short supply chain. On this basis, in the present study the standard quality parameters of semolina and the corresponding pasta obtained from grain selected in different VR-N management zones were evaluated.

2. Materials and Methods

2.1. Field Experiment and Grain Production

A field experiment was carried out at the Miana Serraglia farm (NE Italy, 45°22′ N; 12°08′ E) (Mira, Venice, Italy), located close to the Venetian Lagoon, an area classified with a high risk of nitrate leaching in surface waters and ground waters according to the Nitrate Directive 91/676 [24]. The information about field measures and soil texture were previously reported [2]. Durum wheat (*Triticum durum*

Desf.) var. *Biensur* (Apsovmenti, Voghera, Italy) was grown in 2010–2011 and 2011–2012. Only grain samples harvested in the second year (seeding 24 October 2011 and harvest 4 July 2012) were analyzed in the present work. Weather was characterized by low rainfall and temperature (1.9 °C compared to the 20-year average of 4.1 °C) at the beginning of stem elongation, which prevented early N uptake.

2.2. Selection of Different Management Zones and Variable-Rate Fertilization

A total of 120 samples of the top soil layer (0–30 cm) were collected according to a mixed-sampling scheme [25]. Primary soil properties (texture, bulk density, pH, electric conductivity, organic carbon, total N, and labile phosphorus) were determined according to a previous study [2]. In addition, spatial soil electric conductivity (ECa) was measured with an EMI sensor (Geonics EM38DD). Three management zones (MZs) were delineated, i.e., a high-fertility zone (HFZ), a medium-fertility zone (MFZ), and a low-fertility zone (LFZ). Fertilization with 130, 160, and 200 kg N/ha using ammonium nitrate was applied in HFZ, MFZ, and LFZ, respectively. N doses were defined according to a 30-year model simulation carried out in the three MZs with DSSAT model [26]. Criteria for selecting N doses aimed to balance the productivity with water quality goals (e.g., low N leaching). At the tillering stage, a uniform N rate (52 kg N ha^{-1}) was supplied, while VR-N was managed during stem elongation. Due to the high N amount to be distributed, in LFZ, N application was supplied in two applications to avoid N losses. At the flowering stage, each zone was split into a control (0) and a treatment area, and the latter was treated with a UAN (urea–ammonium–nitrate) solution (15 kg N/ha) (Figure 1). Dates and amount of fertilizations for the 2011–2012 season were previously reported [2].

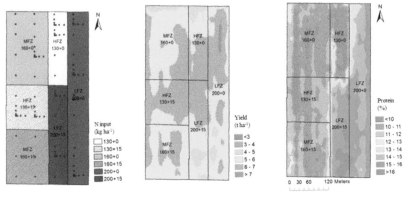

Figure 1. Map of the different fertility zones (**left**), yield (ha; **center**), and protein content (%, **right**) in 2011–2012. In the legend, the N fertilization doses applied were previously reported [2]. HFZ, high-fertility zone, MFZ, medium-fertility zone, LFZ, low-fertility zone. Data derived from reference [2].

Grain yield was recorded by a yield mapping system (Agrocom CL021) mounted on a combine harvester (Claas Lexion 460). Similarly, protein content was measured with a NIR spectrometer that was interfaced to a GPS, as already reported [2]. Crop production ranged from less than 5 t/ha to more than 7 t/ha (Figure 1), with the low productive areas corresponding to sandier areas. N base fertilization was a key determinant of crop yield, but also foliar N application slightly increased grain yield [2]. Protein content fluctuated from less than 10% to more than 15%, mirroring the VR-N areas (Figure 1). In each of the six combinations of MZs (130; 130 + 15; 160; 160 + 15; 200; 200 + 15) and flowering fertilization, a composite grain sample of 100 kg was collected for the chemical composition and pasta analyses.

2.3. Chemical Composition and Gluten Proteins Quantification of Semolina Samples

Grains of the cv. *Biensur* samples (130; 130 + 15; 160; 160 + 15; 200; 200 + 15 kg/ ha) of the 2011–2012 season were ground in an experimental laboratory mill (Buhler MLU202 roller mill; Germany) in order to obtain refined semolina with particle size similar to that of the reference control (200 to 350 μm). The reference control was a commercial semolina (cv. *Aureo*) used for mono-varietal industrial pasta production. The chemical composition of all semolina samples was evaluated according to AOAC standard methods [27] for moisture, protein, starch, fiber, fat, and ash. In addition, the relative quantification of gliadins, high-molecular-weight (HMW) glutenin (GS) fraction, and low-molecular-weight (LMW) glutenin (GS) fraction was carried out using a protein sequential extraction procedure [20] followed by quantification using the Bradford assay [28].

2.4. Dough Analyses and Pasta Production

Dough mixing characteristics (water absorption, dough stability, and dough weakening) were measured in triplicate, by using a Promylograph apparatus equipped with a 100 g bowl (T6, Max Egger, Austria) according to the approved 54-21 method [29].

Short-cut pasta (tubetti) (Figure 2) made from both the *Biensur* samples (130; 130 + 15; 160; 160 + 15; 200; 200 + 15) and reference, was produced using an industrial-scale pilot system at the Pavan-Map Impianti (Galliera Veneta, Padova, Italy). Briefly, pasta samples were prepared in accordance with the Italian legislation [30] by mixing water and semolina to form a dough with a 30% moisture content. The dough was processed using a single-screw extruder (FP 70 model, Pavan) under vacuum conditions. The screw speed was 35 rpm, the cylinder and the extrusion head temperature was 35 °C, while the head pressure was 100 bar. Samples of *Biensur* (130; 130 + 15; 160; 160 + 15; 200; 200 + 15) and reference fresh pasta were transferred to the dryer and treated at decreasing air temperatures (from 90 °C to 45 °C) in a static dryer (SD 100 model, Pavan) to obtain the final moisture content of 11%.

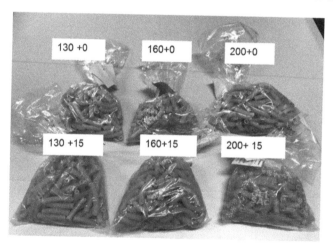

Figure 2. Site-specific pasta manufactured in the management zones.

2.5. Pasta Quality Parameters

Optimal cooking time (OCT), defined as "al dente", was determined by pressing the pasta between two glass slides at different times during cooking in boiling water and observing the time that the starchy white core of the pasta took to disappear [31].

Cooking loss (CL) was evaluated according to method 66-50 [29]. The residue obtained by draining the pasta cooking water was weighted and reported as a percentage of the starting material.

Water absorption (WA) was calculated as the increase in the weight of the pasta after cooking and expressed as a percentage of the weight of the uncooked pasta.

All analyses were carried out with three individual measurements (replicates).

The firmness of cooked pasta was determined using a Texture Analyser (TA.XT plus, Stable Micro Systems, UK) according to 16-50 method [29]. A single tubetto (12 mm thick) was oriented perpendicularly to a knife probe, then compressed at a speed of 0.5 mm/s with a 5 kg load cell. Firmness was measured as the maximum peak force curve (Newton) required to compress the pasta sample. The average value of five replicates was reported.

2.6. Sensory Evaluation of Pasta

To assess the acceptability of the mono-varietal pasta made from cv. *Biensur* obtained from each management zones, a sensory evaluation was carried out by 15 panel members (9 women, 6 men; ages ranging from 22 to 40 years) with experience in food evaluation. The pasta samples (130; 130 + 15; 160; 160 + 15; 200; 200 + 15; reference) were cooked "al dente", drained, and kept warm until serving in randomized order on plastic plates labelled with random two-digit codes. Panelists were asked to evaluate color, flavor, and texture (firmness and stickiness) on a five-point scale from 1, low intensity, to 5, high intensity. They were also asked to score the overall quality of the product based on these same attributes using the same five-point scale. The attribute scores for each sample and panel member were subjected to a one-way analysis of variance (ANOVA) to obtain mean sensory scores for each of the 15 panel members.

2.7. Statistical Analysis

Statistical analysis of the data was performed with the Statgraphics Centurion XIV software (StatPoint Technologies, Inc., Warrenton, VA, USA), and the results were compared with one-way ANOVA. A preliminary Shapiro–Wilk test was applied to test the assumption of normality. Significant differences between treatments were determined by Tukey's test.

3. Results

3.1. Chemical Composition of the Semolina and Dough Analyses

Soil variability and VR-N determined the protein content and composition of the semolina from cv. *Biensur* obtained from different management zones (Table 1).

There was a significant higher protein abundance in grain from LFZ (13.6%, on average) and a lower abundance in grain from HFZ (10.4%, on average), independently of the foliar treatment. The protein content in grain from LFZ was similar to that of the commercial high-quality semolina from cv. *Aureo*, taken as a reference sample (14.7%).

Besides the total protein amounts, by sequential gluten protein extraction, the percentages of the different gluten protein classes (gliadins, HMW-GS, and LMW-GS) were also calculated. These are important parameters affecting dough and pasta technological properties. Gluten strength describes the ability of the proteins to form a tenacious network able to promote better extrusion and superior cooking quality and textural characteristics if compared to weak gluten at the same protein level [32,33]. In particular, several studies showed that adding a glutenin-rich fraction consisting of both HMW-GS and LMW-GS to base semolina, increased the mixograph dough strength and the percentage of unextractable polymeric proteins [33].

Table 1. Proximate composition (expressed as g 100/g of dry matter), gluten proteins quantification, and relative percentage of the three classes of gluten proteins (gliadins (Gli), high-molecular-weight glutenins (HMW-GS) and low-molecular-weight glutenins) (LMW-GS) of site-specific semolina samples of cv. *Aureo*. Data are referred to the 2011–2012 cultivation season.

Soil Fertility Zones	Sample Name *	Total Protein [1] %	Total Gluten Proteins [2] # (mg/g semolina)	% Gli [3]	% HMW-GS [3]	% LMW-GS [3]	HMW/LMW-GS #	Total GS/Gli #	Ash %	Lipids %	Fiber %	Starch %
HFZ	130 + 0	10.2 c	16.9 b	71 a	10 b	24 b	0.4 b	0.4 b	0.74 a	1.85 a	3.07 a	82.68 b
HFZ	130 + 15	10.6 c	16.8 b	69 a	9 b	22 b	0.4 b	0.4 b	0.77 a	1.90 a	2.97 a	82.18 b
MFZ	160 + 0	9.5 d	18.2 b	71 a	9 b	22 b	0.5 b	0.4 b	0.65 a	1.80 a	2.48 b	84.07 a
MFZ	160 + 15	9.2 d	17.8 b	70 a	10 b	17 c	0.6 b	0.3 b	0.69 a	1.79 a	2.74 b	85.16 a
LFZ	200 + 0	13.9 b	24.2 a	65 b	13 b	26 b	0.5 b	0.6 a	0.75 a	1.81 a	3.04 a	79.78 c
LFZ	200 + 15	13.3 b	23.1 a	66 b	12 a	21 b	0.5 b	0.5 a	0.79 a	1.73 a	3.17 a	80.17 c
	Reference	14.7 a	24.2 a	58 c	12 a	30 a	0.4 b	0.7 a	0.78 a	1.80 a	3.01 a	78.86 c

HFZ, high-fertility zone, MFZ, medium-fertility zone, LFZ, low-fertility zone. For each parameter, different letters indicate significant differences (Tukey test, $p \leq 0.05$; $n = 5$). [1] N determined by the Kjeldahl method; [2] Total gluten proteins correspond to the sum of the three different protein fractions (gliadins, HMW-GS, and LMW-GS) extracted according to a published procedure [20]. [3] Percentage of single classes of proteins related to the value of sum of the three gluten protein extracts. * The names of the samples correspond to the N fertilization treatment applied (kg/ha) # Data derived from reference [2].

In addition, it is widely accepted that for the same wheat genotype, the growing environment affects the quality of the semolina. In particular, different works demonstrated that both fertilization dose and fertilization type can modulate the total gluten protein amounts and the relative percentages of different gluten protein fractions (gliadins, HMW-GS, and LMW-GS) both in durum and in common wheat, modifying the HMW-GS/LMW-GS and the glutenin/gliadin (GS/Gli) ratios and consequently the dough strength [34–37].

In this work, we showed that the percentage of gluten proteins was affected by the different soil fertility zones and the fertilization treatments. In particular, HMW-GS increased from the value for the HFZ (9.5% on average with respect to total gluten proteins) to that for LFZ (12.5%, on average with respect to total gluten proteins), thus suggesting an effect of the N treatment on the synthesis of this class of proteins, independently of soil fertility. LMW-GS showed the lower percentage in semolina from MFZ treated with foliar application (17% average with respect to total gluten proteins) while for both LFZ and HFZ, 23.5% of LMW-GS (on average with respect to total gluten proteins) was observed (Table 2), suggesting that their abundance is related to both the soil fertility and the N supplied. Conversely, the percentage of gliadins appeared to be influenced mainly by the fertility of soil, being more abundant for HFZ and MFZ with respect to LFZ (Table 2). As a result of the different gluten protein proportions, the ratio of total GS/Gli increased for LFZ, indicating a higher quality of the gluten protein composition in these semolina samples. Foliar fertilization influenced the percentage and ratio of the different gluten protein fractions only in grain from MFZ.

Table 2. Mixing properties (*n* = 3) of dough samples obtained from site-specific harvest of cv. *Biensur* compared with those of the commercial reference cv. *Aureo*.

Soil Fertility Zones	Sample * Name	Water Absorption (%)	Dough Stability (min)	Dough Weakening (PU)
HFZ	130 + 0	49.3 [c]	7 [b]	65 [a]
HFZ	130 + 15	49.5 [c]	7 [b]	63 [a]
MFZ	160 + 0	48.4 [c]	6 [b]	65 [a]
MFZ	160 + 15	48.0 [c]	6 [b]	68 [a]
LFZ	200 + 0	51.9 [b]	11.0 [a]	32 [b]
LFZ	200 + 15	52.4 [b]	11.5 [a]	31 [b]
	References	55.7 [a]	12.0 [a]	25 [c]

HFZ, high-fertility zone, MFZ, medium-fertility zone, LFZ, low-fertility zone. Within the same column, different letters indicate significant differences (Tukey test, $p \leq 0.05$); * the names of the samples correspond to the N fertilization treatment applied (kg/ha).

We compared all semolina samples for dough stability (length of time the dough maintains its maximum consistency), weakening index (reduction in dough consistency after 20 minutes of mixing), and water absorption (g of water per 100 g of semolina required to reach a dough consistency of 500 PU) (Table 2). The results showed significant differences among cv. *Biensur* samples obtained from different soil fertility zones treated with VR-N (130; 130 + 15; 160; 160 + 15; 200; 200 + 15). In particular, semolina from LFZ, independently from the foliar treatment, showed higher values of dough stability (indicator of the strength) than the other samples, comparable to that of commercial high-quality semolina (>11 min). Samples from LFZ differed also for the amount of water absorbed (ca. 52%) related to protein content (Table 1) and for the dough weakening index (ca. 31), which indicates the tolerance to mechanical mixing. These data indicate that dough properties are also governed by HMW-GS and LMW-GS protein fractions, which are known to be an indicator of strength and coherence of the protein network (Table 1).

3.2. Quality Parameters of Pasta

Pasta quality is expressed in terms of water absorption, cooking loss, and firmness (Table 3). The values of water absorption of all pasta samples at the optimal cooking time were similar (Table 3), corresponding to 170.17–1721.38% of weight increase, in line with what found by other authors [38,39], while the cooking loss, that represent the solid substance leaching in water during cooking, was significantly higher for HFZ and MFZ with respect to LFZ, which presented a cooking loss similar to that of the commercial reference (Table 3). Therefore, the results of the LFZ samples indicate that gelatinized starch is well retained by a strong gluten network that is able to form a compact structure, which was also confirmed by the high firmness value (>5 N), measured by Texture Analyzer. The firmness of the cooked pasta resulted positively correlated to the cooking loss ($r = 0.96$), as indicated by statistical analysis. In general, all standard quality parameters did not reveal any significant differences between LFZ pasta, obtained with or without foliar treatment (200 and 200 + 15) of *Biensur*, and *Aureo* pasta, despite *Biensur* having a lower protein content than *Aureo*, as previously confirmed [20], since it had also a good gluten quality, which plays an important role in the formation of a strong protein network.

Table 3. Pasta quality parameters (optimal cooking time (OCT), water absorption, cooking loss ($n = 3$), and firmness) of pasta from site-specific harvest of cv. *Biensur* compared with those of the commercial reference cv. *Aureo*.

Soil Fertility Zones	Sample Name	Cooking Properties			
		OCT (min.sec)	Water Absorption (%)	Cooking Loss (%)	Firmness (N)
HFZ	130	8.30	170.46 [a]	3.53 [a]	3.83 [b]
HFZ	130 + 15	8.30	170.17 [a]	3.54 [a]	3.85 [b]
MFZ	160	8.30	171.17 [a]	3.94 [a]	3.68 [b]
MFZ	160 + 15	8.30	170.89 [a]	3.94 [a]	3.72 [b]
LFZ	200	9	172.38 [a]	3.0 [b]	5.70 [a]
LFZ	200 + 15	9	171.58 [a]	2.96 [b]	5.69 [a]
	Reference	9	171.90 [a]	2.96 [b]	5.80 [a]

HFZ, high-fertility zone, MFZ, medium-fertility zone, LFZ, low-fertility zone. Within the same column, different letters indicate significant differences (Tukey test, $p \leq 0.05$). * The names of the samples correspond to the N fertilization treatment applied (kg/ha).

The sensory properties of the cooked pasta, such as appearance, flavor, and texture (firmness and stickiness), play an essential role in determining the final consumer preference, especially in traditional pasta-consuming countries [40].

Sensory evaluation of pasta (Figure 3) made from the cv. *Biensur* from each management zones and from the reference cv. *Aureo* showed significant ($p < 0.05$) differences between LFZ pasta (with or without foliar treatment, 200 and 200 + 15) and the others, for all the parameters tested (Figure 3). Finally, pasta obtained from LFZ showed good overall acceptability, comparable to that of the reference pasta. As for the consumer's evaluation on firmness and stickiness, these attributes depend on the intrinsic structure of pasta that is governed by strength and coherence of the gluten network. Therefore, high firmness values correspond to low stickiness values [40,41].

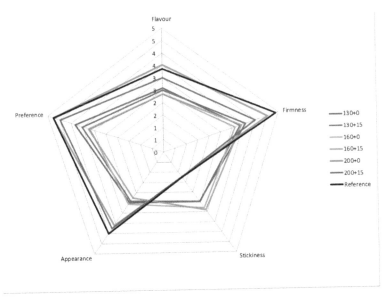

Figure 3. Sensory properties (*n* = 15) of pasta from site-specific harvest of cv. *Biensur* compared with those of the commercial reference cv. *Aureo*. The different fertilization managements listed in the legend are reported in Table 1.

4. Conclusions

In the experiments carried out in this work, the variability of grain protein content in different field zones was significant enough to justify the segregation of grain during harvest. In particular, LFZ, associated with a high N content, showed not only the highest total protein amount in the grain, but also the highest technological quality of gluten proteins.

The higher grain quality in the LFZ was reflected in the elevated dough technological characteristics as well as in the pasta cooking behavior and sensory properties, comparable to those of well-established commercial mono-varietal semolina. Although the experiments demonstrated the feasibility to produce "site-specific" pasta, technical constrains could prevent the application of the described strategy at industrial level. Indeed, a minimum grain stock is required in manufacturing pasta, which could not be achieved if the management zones are limited in size. As observed for precision viticulture [21,23], a multi-field scale approach should be followed to guarantee adequate grain stocks. Within a farm, different fields can be considered at the same time, segregating grain within management zones classified according to crop varieties, pedo-climatic conditions, and management (fertilization, seeding, weed control, etc.). In this way, the manufactured "site-specific pasta" could be a potential asset for a short supply chain, aiming to improve traceability and environmental and economic sustainability.

Author Contributions: Conceptualization, F.M.; Methodology, F.M., G.V., and G.P.; Formal Analysis, G.V. and G.P.; Resources, F.M., G.V., and G.P.; Data Curation, G.V. and G.P.; Writing—Original Draft Preparation, F.M., G.V. and G.P.; Writing—Review and Editing, G.V.; Funding Acquisition, F.M. All authors have read and agreed to the published version of the manuscript.

Funding: Research supported by Project AGER, grant n. 2017-2194.

Conflicts of Interest: The authors declare no conflict of interest.

References

1. Garrido-Lestache, E.; Lopez-Bellido, R.; Lopez-Bellido, L. Durum wheat quality under Mediterranean conditions as affected by N rate, timing and splitting, N form and S fertilisation. *Eur. J. Agron.* **2005**, *23*, 265–278. [CrossRef]
2. Morari, F.; Zanella, V.; Sartori, L.; Visioli, G.; Berzaghi, P.; Mosca, G. Optimising durum wheat cultivation in North Italy. Understanding the effects of site-specific fertilization on yield and protein content. *Precis. Agric.* **2018**, *19*, 257–277. [CrossRef]
3. Dal Ferro, N.; Cocco, E.; Lazzaro, B.; Berti, A.; Morari, F. Assessing the role of agri-environmental measures to enhance the environment in the Veneto Region, Italy, with a model-based approach. *Agric. Ecosys. Environ.* **2016**, *232*, 312–325. [CrossRef]
4. Delin, S. Within-field Variations in Grain Protein Content—Relationships to Yield and Soil Nitrogen and Consistency in Maps Between Years. *Precis. Agric.* **2004**, *5*, 565–577. [CrossRef]
5. Meyer-Aurich, A.; Griffin, T.W.; Herbst, R.; Giebel, A.; Muhammad, N. Spatial econometric analysis of a field-scale site-specific nitrogen fertilizer experiment on wheat (*Triticum aestuvum* L.) yield and quality. *Comp. Electr. Agric.* **2010**, *74*, 73–79. [CrossRef]
6. Karatay, Y.; Meyer-Aurich, A. Profitability and downside risk implications of site-specific nitrogen management with respect to wheat grain quality. *Precis. Agric.* **2019**, *1*, 1–24. [CrossRef]
7. Meyer-Aurich, A.; Weersink, A.; Gandorfer, M.; Wagner, P. Optimal site-specific fertilization and harvesting strategies with respect to crop yield and quality response to nitrogen. *Agric. Syst.* **2010**, *103*, 478–485. [CrossRef]
8. Bongiovanni, R.G.; Robledo, C.W.; Lambert, D.M. Economics of site-specific nitrogen management for protein content in wheat. *Comp. Electr. Agric.* **2007**, *58*, 13–24. [CrossRef]
9. Maertens, K.; Reyns, P.; De Baerdemaeker, J. On-line measurement of grain quality with NR technology. *Trans. ASAE* **2004**, *47*, 1135–1140. [CrossRef]
10. Skerritt, J.H.; Adams, M.L.; Cook, S.E.; Naglis, G. Within-field variation in wheat quality: Implication for precision agricultural management. *Austr. J. Agric. Res.* **2002**, *53*, 1229–1242. [CrossRef]
11. Thylén, L.; Rosenqvist, H. Economical Aspects of sorting grain into different fractions. In *Precision Agriculture. Proceedings of the Sixth International Conference on Precision Agriculture (Conference CD)*; Robert, P.C., Ed.; ASA, CSSA and SSSA: Madison, WI, USA, 2002.
12. Stewart, C.M.; McBratney, A.; Skerrit, J.H. Site-specific durum wheat quality and its relationship to soil properties in a single field in Northern New South Wales. *Precis. Agric.* **2002**, *3*, 155–168. [CrossRef]
13. Martin, C.T.; McCallum, J.D.; Long, D.S. A web-based calculator for estimating the profit potential of grain segregation by protein concentration. *Agron. J.* **2013**, *105*, 721–726. [CrossRef]
14. Tozer, P.R.; Isbister, B.J. Is it economically feasible to harvest by management zone? *Precis. Agric.* **2007**, *8*, 151–159. [CrossRef]
15. Long, D.S.; Engel, R.E.; Siemens, M.C. Measuring grain protein concentration with in-line near infrared reflectance spectroscopy. *Agron. J.* **2008**, *100*, 247–252. [CrossRef]
16. Long, D.S.; McCallum, J.D.; Scharf, P.A. Optical-mechanical system for on-combine segregation of wheat by grain protein concentration. *Agron. J.* **2013**, *105*, 1529–1535. [CrossRef]
17. Gandorfer, M.; Meyer Aurich, A. Economic potential of site-specific fertilizer application and harvest management. In *Precision Agriculture: Technology and Economic Perspectives*; Pedersen, S.M., Lind, K.M., Eds.; Springer International Publishing: Basel, Switzerland, 2017; pp. 79–92.
18. Bevilacqua, M.; Braglia, M.; Carmignani, G.; Zammori, F.A. Life cycle assessment of pasta production in Italy. *J. Food Qual.* **2007**, *30*, 932–952. [CrossRef]
19. Italian Legislative Decree Indicazione Dell'origine, in Etichetta, del Grano duro per Paste di Semola di Grano duro (17A05704). Gazzetta Ufficiale n. 191. p. 16. Available online: https://www.gazzettaufficiale.it/eli/id/2017/08/17/17A05704/sg (accessed on 17 August 2017).
20. Visioli, G.; Vamerali, T.; Dal Cortivo, C.; Trevisan, S.; Simonato, B.; Pasini, G. Pasta-making properties of the new durum wheat variety *Biensur* suitable for the northern Mediterranean environment. *It. J. Food Sci.* **2018**, *30*, 673–683.
21. Bramley, R.; Pearse, B.; Chamberlain, P. Being Profitable Precisely: A case study of Precision Viticulture from Margaret River. *Aust. New Zealand Grapegrow. Winemak.—Annu. Tech. Issue* **2003**, *473a*, 84–87.

Foods **2020**, *9*, 477

22. Santesteban, L.G.; Guillaume, S.; Royo, J.B.; Tisseyre, B. Are precision agriculture tools and methods relevant at the whole-vineyard scale? *Precis. Agric.* **2013**, *14*, 2–17. [CrossRef]

23. Bramley, R.G.V.; Ouzman, J.; Thornton, C. Selective harvesting is a feasible and profitable strategy even when grape and wine production is geared towards large fermentation volumes. *Austr. J. Grape and Wine Res.* **2011**, *17*, 298–305. [CrossRef]

24. EC-Council Directive. Council Directive 91/676/EEC Concerning the Protection of Waters Against Pollution Caused by Nitrates from Agricultural Sources. 1991. Available online: https://eur-lex.europa.eu/eli/dir/1991/676/2008-12-11 (accessed on 11 December 2008).

25. Morari, F.; Loddo, S.; Berzaghi, P.; Ferlito, J.C.; Berti, A.; Sartori, L.; Visioli, G.; Marmiroli, N.; Piragnolo, D.; Mosca, G. Understanding the effects of site-specific fertilization on yield and protein content in durum wheat. In *Precision Agriculture '13 –Proceedings of the Nineth European Conference on Precision Agriculture*; Stafford, J.V., Ed.; Wageningen Academic Publishers: Wageningen, The Netherlands, 2013; pp. 321–327.

26. Sartori, L. Sud-Project 3. Precision agriculture and conservtion agriculture. Action 1: Precision agriculture. In *Identification of Innovative Cropping Systems in Venice Lagoon Watershed. Summary Results*; Veneto Agricoltura (In Italian): Padova, Italy, 2010; pp. 67–76.

27. Association of Official Analytical Chemists. *Official Methods of Analysis*, 17th ed.; AOAC, Inc.: Arlington, VA, USA, 2000.

28. Bradford, M. A rapid and sensitive method for the quantitation of microgram quantities of protein utilizing the principle of protein-dye binding. *Anal. Biochem.* **1976**, *72*, 248–254. [CrossRef]

29. American Association of Cereal Chemists. *Approved Methods of the AACC*, 10th ed.; Methods 44-15A, 66-50 and 985.29; American Association of Cereal Chemists: St. Paul, MN, USA, 2000.

30. Presidential Decree n.187, 2001. Regolamento per la Revisione Della Normativa Sulla Produzione e Commercializzazione di Sfarinati e Paste Alimentari, a Norma Dell'articolo 50 Della Legge 22 Febbraio 1994, n. 146. Available online: http://www.edizionieuropee.it/LAW/HTML/24/zn4_03_057.html (accessed on 22 May 2001).

31. Abécassis, J.; Abbou, R.; Chaurand, M.; Morel, M.H.; Vernoux, P. Influence of extrusion conditions on extrusion speed, temperature, and pressure in the extruder and on pasta quality. *Cereal Chem.* **1994**, *71*, 247–253.

32. Simmonds, D.H. *Wheat and Wheat Quality in Australia*; CSIRO: Canberra, Australia, 1989.

33. Sissons, M. Role of Durum Wheat Composition on the Quality of Pasta and Bread. *Food* **2008**, *2*, 75–90.

34. Visioli, G.; Bonas, U.; Cortivo, C.D.; Pasini, G.; Marmiroli, N.; Mosca, G.; Vamerali, T. Variations in yield and gluten proteins in durum wheat varieties under late-season foliar versus soil application of nitrogen fertilizer in a northern Mediterranean environment. *J. Sci. Food Agric.* **2018**, *98*, 2360–2369. [CrossRef] [PubMed]

35. Dupont, F.M.; Hurkman, W.J.; Vensel, W.H.; Tanaka, C.; Kothari, K.M.; Chung, O.K.; Altenbach, S.B. Protein accumulation and composition in wheat grains: Effects of mineral nutrients and high temperature. *Eur. J. Agron.* **2006**, *25*, 96–107. [CrossRef]

36. Wan, Y.; Shewry, P.R.; Hawkesford, M.J. A novel family of γ-gliadin genes are highly regulated by nitrogen supply in developing wheat grain. *J. Exp. Bot.* **2013**, *64*, 161–168. [CrossRef]

37. Galieni, A.; Stagnari, F.; Visioli, G.; Marmiroli, N.; Speca, A.; Angelozzi, G.; D'Egidio, S.; Pisante, M. Nitrogen fertilisation of durum wheat: A case study in Mediterranean area during transition to conservation agriculture. *It. J. Agron.* **2016**, *662*, 12–23. [CrossRef]

38. Bruneel, C.; Pareyt, B.; Brijs, K.; Delcour, J.A. The impact of the protein network on the pasting and cooking properties of dry pasta products. *Food Chem.* **2010**, *120*, 371–378. [CrossRef]

39. Padalino, L.; Mastromatteo, M.; Lecce, L.; Spinelli, S.; Contò, F.; Del Nobile, M.A. Effect of durum wheat cultivars on physico-chemical and sensory properties of spaghetti. *J. Sci. Food Agric.* **2014**, *11*, 2196. [CrossRef]

40. D'Egidio, M.G.; Nardi, S. Textural measurement of cooked spaghetti. In *Pasta Noodle Technology*; Kruger, J.E., Matsuo, R.B., Dick, J.W., Eds.; AACC: Washington, DC, USA, 1998; pp. 133–169.

41. Cubadda, R.E.; Carcea, M.; Marconi, E.; Trivisonno, M.C. Influence of gluten proteins and drying temperature on the cooking quality of durum wheat pasta. *Cereal Chem.* **2007**, *84*, 48–55. [CrossRef]

Article

Cooking Effect on the Bioactive Compounds, Texture, and Color Properties of Cold-Extruded Rice/Bean-Based Pasta Supplemented with Whole Carob Fruit

Claudia Arribas [1], Blanca Cabellos [1], Carmen Cuadrado [1], Eva Guillamón [2] and Mercedes M. Pedrosa [1,*]

1   Food Technology Department, SGIT-INIA, Ctra de La Coruña, Km 7.5, 28040 Madrid, Spain;
    correoinia@gmail.com (C.A.); blancacc@inia.es (B.C.); cuadrado@inia.es (C.C.)
2   Centre for the Food Quality, SGIT-INIA, C/Universidad s/n, 42004 Soria, Spain; guillamon.eva@inia.es
*   Correspondence: mmartin@inia.es

Received: 3 March 2020; Accepted: 30 March 2020; Published: 2 April 2020

Abstract: Pasta is considered as the ideal vehicle for fortification; thus, different formulations of gluten-free pasta have been developed (rice 0–100%, bean 0–100%, and carob fruit 0% or 10%). In this article, the content of individual inositol phosphates, soluble sugars and α-galactosides, protease inhibitors, lectin, phenolic composition, color, and texture were determined in uncooked and cooked pasta. The highest total inositol phosphates and protease inhibitors contents were found in the samples with a higher bean percentage. After cooking, the content of total inositol phosphates ranged from 2.12 to 7.97 mg/g (phytic acid or inositol hexaphosphate (IP6) was the major isoform found); the protease inhibitor activities showed values up to 12.12 trypsin inhibitor (TIU)/mg and 16.62 chymotrypsin inhibitor (CIU)/mg, whereas the competitive enzyme-linked immunosorbent assay (ELISA) showed the elimination of lectins. Considering the different α-galactosides analyzed, their content was reduced up to 70% ($p < 0.05$) by the cooking process. The total phenols content was reduced around 17–48% after cooking. The cooked samples fortified with 10% carob fruit resulted in darker fettuccine with good firmness and hardness and higher antioxidant activity, sucrose, and total phenols content than the corresponding counterparts without this flour. All of the experimental fettuccine can be considered as functional and healthy pasta mainly due to their bioactive compound content, compared to the commercial rice pasta.

Keywords: prebiotics; trypsin inhibitors; inositol phosphates; phenols; legumes; functional foods; gluten-free

1. Introduction

Pasta is largely consumed all over the world and, moreover, these kinds of products are considered by the World Health Organization (WHO) as an ideal vehicle for fortification [1]. Pasta is a cereal-based product characterized by good organoleptic and nutritional properties and long shelf life. Pasta, such as spaghetti and macaroni, are manufactured traditionally by mixing durum wheat semolina and water (and eggs in some types of products). During the wheat pasta-making process, gluten is the principal compound responsible for the formation of the pasta structure, and it provides better quality parameters (low loss of solids in the cooking process, firm structure, attractive color, etc.) than those of gluten-free (GF) pasta mainly elaborated with corn, sorghum, rice, and/or legumes [2]. Nowadays, the population with celiac disease, as well as those who exclude the gluten-containing traditional products from their diet for other health reasons, is growing; therefore, there is a strong need for the development of novel nutritious and healthy products.

In order to elaborate GF pasta, the gluten matrix of the traditional pasta products can be substituted by other starchy raw materials and/or new formulations, alternative pasta-making processes, and/or the addition of additives; the raw materials and the pasta-making process have a great effect on the quality of the pasta [3]. The use of starchy cereals, such as rice, combined with pulses can provide nutritive and balanced pasta products, with a good content and profile of proteins, dietary fiber, complex carbohydrates, minerals, and vitamins [4,5]. As reported by Elliott et al. [6], GF products are not nutritionally finer than gluten-containing products. Thus, these new pasta products based on cereal and pulse mixtures would be of great interest to celiac patients, as well as to those people that decide to eliminate gluten from their diets. White rice is the main ingredient used by the food industry to manufacture GF products, even though it is a material with a poor bioactive compound content. On the contrary, pulses are rich in bioactive compounds (e.g., phenolic compounds, phytates, saponins, α-galactosides, or trypsin inhibitors). Traditionally, some of these phytochemicals such as α-galactosides, trypsin inhibitors, or phytates have been considered as anti-nutritive compounds that can produce flatulence, reduce protein digestion, or impair the absorption of other compounds, respectively. However, at present, it is well known that these same compounds are related to a reduction in the risk of cardiovascular disease, type 2 diabetes, digestive tract diseases, becoming overweight, and obesity [5]. Therefore, the fortification of rice-based pasta with legumes would increase the amount of bioactive compounds, allowing us to obtain nutritious and healthy foods. Among the pulses, beans (*Phaseolus vulgaris* L.) are consumed worldwide and, moreover, with the growing of vegetarianism and the demand in Western countries for non-wheat and non-soy proteins, dry beans are obtaining increased attention by consumers; nevertheless, they are underutilized by the food industry as an ingredient in novel foods. Although, in general, all beans have a similar nutritive composition, each variety has a unique bioactive compound profile. In this study, the Almonga variety was used, which was previously reported to have a low lectin content, and whose consumption results in a significant decrease in triglyceride levels [7]. Considering that *Phaseolus vulgaris* lectin (PHA) is the main toxic compound present in beans, the use of this variety could be of great interest in the production of safe and healthy foods, notably, in the formulation of pasta products that require short cooking time. Some authors have developed rice-based pasta enriched with different pulses, although these pastas showed poor cooking quality, mainly due to a reduction in the integrity of the protein matrix [2,8].

A legume with high potential to be used as an ingredient to improve the quality of GF pasta is carob (*Ceratonia siliqua* L.). Carob seeds are utilized for the production of a thickener that is used in the food industry called carob bean or locus bean gum (E-410) [9,10]. Additionally, carob seeds contain a protein called caroubin, which has similar characteristics to wheat gluten but is safe for celiac people. Carob fruits are also rich in dietary fiber polyphenols and contain moderate amounts of inositol phosphates and α-galactosides [9,11]. The high amount of dietary fiber in carob fruit can weaken the firmness of the pasta, since it can disturb the protein matrix. Biernacka et al. enriched wheat pasta with up to 4% carob fiber with little effect on pasta quality and with good overall acceptability [12]. Therefore, the inclusion of a limited amount of whole carob fruit (WCF; pod and seeds) can be a promising ingredient in the development of good quality GF pasta and can obtain functional products [13] due to the presence of locus bean gum and caroubin in the seeds, as well as the presence of phenolic compounds in the pods. Some authors have reported the use of kibble, carob germ proteins, or carob bean gum in some bread or pasta products [12,14,15]. Turfani et al. [9] elaborated bread with lentil and carob seeds (up to 12%). In other previous works [4,11], the addition of WCF (5–10%) to extruded 'ready-to-eat' GF snacks based on rice/bean or rice/pea has created foods with a good bioactive compound content able to promote health functions and to improve the textural attributes of the snacks. Although, to the best of our knowledge, there are no studies regarding the development of novel pasta products containing both WCF plus another legume.

It is important to note that the content of bioactive compounds in any end-food product depends on the processing technique utilized in their development (such as dehulling, canning, soaking, germination, fermentation, autoclaving, extrusion-cooking, cold extrusion, etc.). Moreover, considering that pasta is consumed after cooking and that this process can induce changes in the phytochemical content, mainly in the heat-labile compounds [16,17], it is essential to know the real amount (i.e., that which would be consumed) present in the pasta after cooking.

Taking into account the above information, the aim of this study was to determine the content of α-galactosides, mio-inositol phosphates, protease inhibitors, lectins, and phenols in dry (uncooked) and cooked experimental samples based on different proportions of rice/bean and supplemented with WCF, as well as in a commercial rice-based pasta whose phytochemical content has not been previously studied. The pasta quality was assessed by determining the texture and color properties.

2. Materials and Methods

2.1. Raw Materials

White rice (*Oryza sativa* L.), raw beans (*P. vulgaris* var. Almonga), and whole carob fruit (WCF; pod and seeds) (*C. siliqua* L.) were the ingredients used to develop the GF pasta. The raw materials were acquired from Cámara Arrocera de Amposta (Tarragona, Spain), ITACyL (Instituto Tecnológico Agrario de Castilla y León, Valladolid, Spain), and Armengol Hermanos (Tarragona, Spain), respectively. All raw materials were milled and passed through a 1 mm sieve (Retsch SK1, Haan, Germany), and then stored in polyethylene bags until pasta elaboration. In order to compare the experimental pasta with a commercial product, a commercial fresh rice pasta was purchased from a local market.

2.2. Pasta Formulations and Manufacturing

Ten pasta formulations were elaborated by mixing different ratios of rice and bean, and supplemented with 10% whole carob bean (Table 1). The pasta was elaborated following the methodology (with minor modifications) of Gallegos-Infante et al. and Giuberti et al. [8,18]. The total amount of each formulation processed was 1 kg. To elaborate the fettuccine-shaped pasta, the different formulations were mixed (5 min) with an appropriate amount of hot water (46% on average) in a domestic blender (Thermomix TM-31, Vorweck, Wuppertal, Germany) to allow a uniform distribution of the water throughout the flour particles. The hydrated flours were worked in the mixing chamber of a continuous pilot-scale extruder for pasta production (Imperia & Monferrina S.p.A., Dolly, Moncalieri, Italy) for 15 min at constant speed to obtain aggregates of a 3–5 mm diameter. Afterward, a conventional cold (30–40 °C) extrusion was carried out in the same continuous pilot-scale extruder to produce experimental fettuccine (20 cm in length). The pasta was pre-dried at room temperature (30 min) and then in an oven at 70 °C (around 2 h). The commercial fresh rice pasta was dried in the same way as the experimental samples (Figure S1). Five hundred grams of each dried pasta was milled and passed through a 1 mm sieve, and then stored in polyethylene bags. The other 500 g of dried pasta was stored at room temperature in polyethylene boxes until required for the cooking procedure and the texture and color determinations.

The dry pasta (P-) was elaborated and the commercial sample was cooked (PC-) at the optimal cooking time (OCT). The OCT was analyzed in a previous paper, and was 4 min for the P-Commercial rice, 2 min for P-Rice, and for the rest of the experimental fettuccine, it ranged from 2.6 min for P-20.0 to 3.92 min for P-80.10 [19]. All of the cooked samples were frozen immediately (−20 °C), lyophilized, and stored in polyethylene bags until required; then, they were ground in a mill equipped with a 1 mm sieve (Retsch SK1 mill, Haan, Germany).

Table 1. Codification and formulation of the different rice/bean-based fettuccine pastas fortified with carob fruit flours and the commercial control pasta (external control).

Formulation	Rice (%)	Bean (%)	Carob Fruit (%)
20.0	80	20	0
20.10	70	20	10
40.0	60	40	0
40.10	50	40	10
60.0	40	60	0
60.10	30	60	10
80.0	20	80	0
80.10	10	80	10
Bean 100%	0	100	0
Rice 100%	100	0	0
Commercial rice	The product label: rice flour, corn flour, thickener and emulsifier additives, and water.		

Note: All of the experimental samples contained 1% of sodium chloride in the formulation.

2.3. Biochemical Characterization

All of the biochemical analyses were performed in the uncooked (P-) and the cooked samples (PC-).

2.3.1. Individual Inositol Phosphates

The individual phosphates, from inositol triphosphate (IP3) to phytic acid (IP6), were determined according to the method of Burbano et al. [20]. Sodium phytate was used as the standard (Sigma-Aldrich, St. Louis, MO, USA). The samples were analyzed using high performance liquid chromatography (HPLC) (Beckman System Gold Instrument, Los Angeles, CA, USA) equipped with a refractive index detector and a macroporous polymer PRP-1 column (150 × 4.1 mm i.d., 5 µm, Hamilton, Reno, Nevada, USA) maintained at 45 °C with a flow rate of 1 mL/min. The mobile phase consisted of a mixture of methanol/water (52/48 *v/v*) plus 8 mL tetrabutylammonium hydroxide (40% in water), 1 mL 5 M sulphuric acid, 0.5 mL 91% formic acid, and 100 µL of phytic acid (6 mg/mL), and the pH was adjusted to 4.3.

2.3.2. Soluble Sugars and α-Galactosides

The content of soluble sugars and α-galactosides was analyzed by HPLC (Beckman System Gold Instrument, Los Angeles, CA, USA) with a refractive index detector according to Pedrosa et al. [21]. The sugars present in each sample (0.1 g) were extracted with ethanol/water (50/50, *v/v*). Then, the extract was purified using Sep-Pack C18 cartridges (500 mg, Waters, Milford, MA, USA) and injected into the HPLC system. The mobile phase consisted of acetonitrile/water 60:40 (*v/v*) and was used in isocratic mode to equilibrate the column (Spherisorb-5-NH2, 250 × 4.66 mm i.d. Waters, Milford, MA, USA) at a 1 mL/min flow rate. The content of individual sugars was quantified by comparison to their corresponding standards (Sigma-Aldrich, St. Louis, MO, USA). Ciceritol was purified and kindly supplied by Dr. A. I. Piotrowicz-Cieslak (Olsztyn-Kortowo, Poland).

2.3.3. Trypsin and Chymotrypsin Inhibitors

The trypsin (TI) and chymotrypsin (CI) inhibitors were extracted according to Pedrosa et al. [21]. α-N-benzoyl-DL-arginine-p-nitroanilide hydrochloride (BAPNA) and benzoyl-L-tyrosine ethyl ester (BTEE) were used as substrate of the trypsin and the chymotrypsin, respectively. Trypsin inhibitor activity was determined at 410 nm, and one unit of trypsin inhibitor (TIU) per milligram of flour was defined as a decrease of 0.01 units of absorbance at 410 nm in relation to the trypsin control, using a 10 mL assay volume [22]. Chymotrypsin inhibitor activity was determined at 256 nm. One unit of chymotrypsin inhibitor (CIU) was stated as an increase of 0.01 absorbance units at 256 nm after 7 min after the addition of the substrate to the reaction mixture.

2.3.4. Lectin

The *Phaseolus vulgaris* lectin (PHA) content was determined using a non-commercial competitive indirect ELISA assay according to Cuadrado et al. [23]. The PHA content of the samples was calculated using a calibration curve (0.001–1000 µg/mL) of pure PHA standard. The limit of detection and of quantification were estimated from the regression equation calculated from the calibration curve ($y = -0.4311Ln(x) + 2.711$; $r^2 = 0.98$). *P. vulgaris* cvs. Processor and Pinto were included in each assay as positive and negative controls, respectively, to determine the repeatability of the method. The results are expressed as percentage of PHA on a dry matter basis.

2.3.5. Phenolic Composition and Antioxidant Activity

A solution of methanol–HCl (1‰)/water (80:20 *v/v*) was used to extract the phenolic compounds of the different samples for 16 h at room temperature [24]. The different groups of phenols present in the extracts were quantified spectrophotometrically following the methodology of Oomah et al. [25]. Anthocyanins, flavonols, tartaric esters, and the total phenolic compounds were monitored at 520, 360, 320, and 280 nm, respectively, on a Beckman spectrofotometer (Beckman DU-640, Los Angeles, CA, USA). Commercial standards of cyanidin-3-glucoside from Extrasynthese (Germay, France), quercetin, caffeic acid, and catechin from Sigma-Aldrich (St. Louis, MO, USA) were used to quantify anthocyanins (µg of cyanidin-3-glucoside equivalents (C3GE) per gram dry weight), flavonols (µg of quercetin equivalents (QE) per gram of dry weight), tartaric esters (mg of caffeic acid equivalents (CAE) per gram of dry weight) and total phenols (mg of (+) catechin equivalents (CE) per gram of dry weight), respectively. The same extracts were used to determine the antioxidant activity by using the oxygen radical absorbance capacity (ORAC) assay [11]. A microplate fluorescence reader (FLUOstar Omega, BMG Labtech, Offenburg, Germany) was used with excitation and emission wavelengths at 485 and 530 nm, respectively. The ORAC values were calculated from a calibration curve of Trolox (0–8 µg/mL) and the results were expressed as µmol Trolox per gram of dry weight).

2.4. Color Analysis

The color of the uncooked and the cooked fettuccine pastas in the optimum cooking time was measured using the Konica Minolta model CM-5 spectrophotometer (Konica, Minolta, Ramsey, NJ, USA) with a xenon lamp. The results are expressed in the color space CIE L*a*b*, where L* measures the degree of luminosity (lightness/darkness), a* (redness/greenness) where +a* are red tones and -a*are green tones, and b* (yellowness/blueness), where +b* yellow tones and -b* blue. The results are the mean of three replicates per sample, measured in each replica per quadruplicate [26]. The calibration (zero and white) was carried out (with zero calibration box and an internal white plate, respectively) and used to standardize daily the equipment before the measurements.

2.5. Texture Analysis

The evaluation of the texture [27] of the cooked (OCT) pasta was done using a texturometer (Model TA-XTplus, Stable Micro System, Surrey, UK) equipped with a load cell of 30 kg coupled to an aluminum P25 mm probe and calibrated daily using a 5 kg load cell. The instrument settings were as follows: compression mode with force pre-set, test, and post-test speed of 2 mm/s and deformation of 75%. The parameters measured were: hardness (force maximum positive), stickiness (force maximum negative), and adhesiveness (negative area). In addition, a cutting test was carried out with a Warner Bratzler at a speed of 0.17 mm/min, being the parameters determined: firmness (force maximum) and consistency (area up to the breaking point of the pasta).

All of the measurements were carried out after cooling (1 min) the samples in cool water in order to end the cooking process. The results are expressed as the mean of four measurements from three different cooking replications of each sample. Texturometer software (Exponent Stable Micro System version 5.0.9.0; Stable Micro System, Surrey, UK) was used to record and calculate the values of the parameters analyzed.

2.6. Statistical Analysis

Results are presented as mean ± standard deviation (SD). They were obtained in quadruplicate, except for color and texture analysis ($n = 12$). The statistically significant differences ($p < 0.05$) were established by a one-way ANOVA, and a Duncan's multiple range test was applied. Correlations were analyzed by Pearson's test. In addition, data from the different chemical analyses were subjected to a principal component analysis (PCA), shown in the supplementary files (Figures S2 and S3). The Statgraphics Centurion XVI computer package (Graphics Software System, Rockville, MD, USA) was used.

3. Results and Discussion

The data corresponding to the different bioactive compounds present in the raw samples are shown in the corresponding tables (Tables 2–5). Considering these results (on a dry basis), raw bean contained the highest amount of total inositol phosphate (IP), α-galactosides, and protease inhibitors, while WCF presented the highest content of sucrose and phenolic compounds. Raw bean revealed very low amounts of lectins, while rice presented low amounts of sucrose and trypsin inhibitors. The content of bioactive compounds determined in these raw materials was similar to that reported in the literature [11,12,16,28].

3.1. Individual Inositol Phosphates

The total and individual inositol phosphates (IP) content in the uncooked (P-) and cooked (PC-) fettuccine, as well as in the commercial rice pasta, are presented in Table 2.

Table 2. Inositol phosphates content (mg/g dry weight) of the raw materials and the different uncooked (P-) and cooked (PC-) fettuccine pastas and the commercial rice pasta.

Sample	IP3	IP4	IP5	IP6	Total Inositol Phosphates
Bean	0.26 ± 0.01	0.42 ± 0.01	1.39 ± 0.03	10.12 ± 0.03	12.20
Carob fruit	n.d.	0.15 ± 0.01	0.36 ± 0.04	0.15 ± 0.01	0.66
Rice	0.10 ± 0.01	0.03 ± 0.03	0.22 ± 0.01	1.53 ± 0.05	1.88
P-20.0	0.23 ± 0.001 [e f, A]	0.39 ± 0.02 [b, A]	0.71 ± 0.07 [b, A]	2.51 ± 0.13 [c, A]	3.83 [c, A]
P-20.10	0.22 ± 0.001 [c d, A]	0.36 ± 0.02 [b, A]	0.70 ± 0.04 [b, A]	2.51 ± 0.18 [c, A]	3.79 [c, A]
P-40.0	0.25 ± 0.02 [j, A]	0.55 ± 0.03 [d e A]	1.08 ± 0.04 [c d, A]	3.29 ± 0.21 [e f, A]	5.16 [d, A]
P-40.10	0.22 ± 0.001 [d e, A]	0.52 ± 0.05 [c d, A]	0.97 ± 0.18 [c, A]	3.11 ± 0.12 [e, A]	4.83 [d, A]
P-60.0	0.24 ± 0.01 [h i, A]	0.56 ± 0.02 [c d, A]	1.28 ± 0.07 [d e, A]	3.57 ± 0.11 [f g, A]	5.65 [e, A]
P-60.10	0.24 ± 0.01 [g h i, A]	0.64 ± 0.10 [f, A]	1.18 ± 0.57 [c d e, A]	4.09 ± 0.34 [h, A]	6.15 [f, A]
P-80.0	0.23 ± 0.01 [f g, A]	0.66 ± 0.03 [f g, A]	1.84 ± 0.08 [f g, A]	4.80 ± 0.11 [j, A]	7.53 [h, A]
P-80.10	0.24 ± 0.01 [i j, A]	0.70 ± 0.01 [g, A]	1.97 ± 0.09 [g, A]	5.16 ± 0.20 [j, A]	8.06 [i, A]
P-Bean	0.25 ± 0.001 [h, A]	0.75 ± 0.02 [h, A]	2.29 ± 0.12 [h, A]	5.73 ± 0.58 [k, A]	9.03 [j, A]
P-Rice	n.d.	0.27 ± 0.02 [a, A]	0.47 ± 0.12 [a, A]	1.48 ± 0.13 [b, A]	2.22 [b, A]
P-Commercial rice	n.d.	0.25 ± 0.01 [a, A]	0.29 ± 0.01 [a, A]	0.37 ± 0.06 [a, A]	0.84 [a, A]
PC-20.0	0.22 ± 0.001 [b c d, B]	0.39 ± 0.02 [b, A]	0.74 ± 0.03 [b, A]	2.81 ± 0.20 [d, B]	4.15 [c, A]
PC-20.10	0.21 ± 0.001 [b, B]	0.38 ± 0.01 [b, A]	0.76 ± 0.02 [b, A]	2.40 ± 0.18 [c, A]	3.75 [c, A]
PC-40.0	0.22 ± 0.001 [c d, B]	0.48 ± 0.02 [c, B]	1.07 ± 0.07 [c d, A]	3.35 ± 0.08 [e f, A]	5.12 [d, A]
PC-40.10	0.21 ± 0.001 [b c d, A]	0.48 ± 0.03 [c, B]	1.18 ± 0.11 [c d e, A]	3.35 ± 0.06 [e f, A]	5.23 [d, A]
PC-60.0	0.23 ± 0.001 [e f, B]	0.52 ± 0.03 [c d e, A]	1.27 ± 0.10 [d e, A]	3.84 ± 0.12 [g h, A]	5.86 [e f, A]
PC-60.10	0.22 ± 0.001 [d e, B]	0.54 ± 0.03 [d e, B]	1.39 ± 0.06 [e, A]	3.91 ± 0.17 [h, A]	6.05 [f, A]
PC-80.0	0.23 ± 0.001 [f g h, A]	0.63 ± 0.04 [f, A]	1.76 ± 0.12 [f g, A]	4.82 ± 0.16 [j, A]	7.43 [h, A]
PC-80.10	0.23 ± 0.001 [f g, B]	0.57 ± 0.01 [e, B]	1.64 ± 0.05 [f, B]	4.54 ± 0.32 [j, B]	6.98 [g, B]
PC-Bean	0.21 ± 0.001 [b c, B]	0.57 ± 0.03 [c d, A]	1.91 ± 0.08 [g, B]	5.28 ± 0.11 [j, B]	7.97 [i, A]
PC-Rice	n.d.	0.26 ± 0.01 [a, A]	0.46 ± 0.03 [a, A]	1.40 ± 0.08 [b, A]	2.12 [b, A]
PC-Commercial rice	n.d.	0.29 ± 0.01 [a, A]	0.34 ± 0.03 [a, A]	0.53 ± 0.12 [a, A]	1.16 [a, A]

Values are means ± standard deviation ($n = 4$). Mean values in the same column followed by a different superscript letter are significantly different ($p < 0.05$); small superscript letters mean differences among all of the samples analyzed, whereas capital superscript letters mean differences due to the cooking treatment for the same formulation. n.d., not detected.

As expected, a higher total IP content corresponded to the samples with a higher bean percentage. A linear correlation between the percentage of bean flour in the formulas and the total inositol phosphates content in uncooked and cooked samples was fitted according to the model: Total IP (mg/g) = 2.071 + 0.067 × %bean, and it showed a high correlation (R^2 = 0.96).

In the experimental dry or uncooked fettuccine, P-Bean showed the highest total IP content (9.03 mg/g), and P-Commercial rice pasta presented the lowest value (0.84 mg/g) (Table 2). On the other hand, the IP content in the cooked samples ranged between 1.16 and 7.97 mg/g for the PC-Commercial rice pasta and PC-Bean, respectively. There were no significant differences in the total IP content between the uncooked and cooked samples (except for the 80.10 formula), probably due to the heat-stable characteristics of phytic acid [29]. In general, the total inositol content showed a slight increase, which was not statistically significant, in the cooked samples (PC-). This could be due to the fact that during pasta-making, some inositol can form complexes with other food components (e.g., minerals, proteins, or starch), which would be released from the pasta matrix during cooking [30]. This slight increase could also be related to the losses of some components easily soluble in water, such as sugars or phenols (Tables 3 and 5), that migrate to the cooking solution and that are discarded; this reduces the total dry matter of the samples and causes changes in the percentage of the specific components expressed on a dry matter basis [31]. The obtained results were in concordance with those published by Tazrart et al. [32] in maccheronccini elaborated with durum wheat semolina fortified with different ratios of faba bean flour (*Vicia faba* L.). In contrast, soaking yellow field peas (*Pisum sativum* L.) revealed that the levels of IP6 were reduced, and, also, the canning process of 'ready-to-eat' Spanish beans decreased significantly the total IP content [17].

These differences could be due to the differences in the cooking time and the temperature used to process the pulses. IP6 was the main form determined in all of the uncooked and cooked samples studied, accounting for, in general, about 65–75% of the total IP content, similar to the results published for different legumes by Burbano et al. [20]. In comparison to the commercial sample, the experimental fettuccine showed from 4 to 15 times more IP6 content, which is in concordance with the results published by Bilgicli et al. [33] for GF noodles fortified with legume (chickpea or soya). The same tendency was shown in semolina-based pasta enriched by α-galactosides-free lupin flours [34].

The amount of the less-phosphorylated IP forms (IP3–IP5) was similar in the uncooked and cooked samples. Therefore, the thermal process did not produce significant differences in the IP isoform content of most of the samples, except in the formulations with a high bean content and WCF. On the other hand, the IP3 content represented about 2.55–6.15% of the total IP content, being detected in all of the samples except in the uncooked and cooked rice 100% and the commercial rice samples (Table 2).

Although, phytic acid (IP6) can reduce the availability of some minerals and, at present, there is not a recommended dietary intake of phytate. Fredrikson et al. (2001) [35] reported that a content of total IP < 20 mg/g (as in the experimental fettuccine) allows a similar degree of mineral availability for commercial food based on soybean protein. Phytic acid has been linked to the epidemiological relation between the high-fiber diets (rich in IP6) and the low incidence of some cancers [11,16]. Also, it can reduce the toxicity of heavy metals (such as lead and cadmium) present in the diet. The less-phosphorylated isoforms (IP3–IP5) have been shown to be related to beneficial effects in human health, such as improving the hypercholesterolemia and the atherosclerosis, and preventing the formation of kidney stones, colon cancer, type 2 diabetes, and irritable bowel syndrome. The results obtained in the experimental fettuccine suggests that the cooking process was not strong enough to significantly affect the contents of phytic acid and the less-phosphorylated forms (IP5–IP3); consequently, they could retain the health benefits associated with the less-phosphorylated inositol phosphates.

3.2. Soluble Sugars and α-Galactosides

The content of the different soluble sugars detected in the uncooked (P-) and cooked (PC-) fettuccine, as well as in the commercial rice pasta, are presented in Table 3. In the current study, sucrose was detected in all of the formulations, and significant differences ($p < 0.05$) between uncooked and cooked counterparts were observed. P-Rice was the sample with highest sucrose content (64.46 mg/g), whereas PC-20.0 showed the lowest amount (5.76 mg/g). As WCF had a high content of sucrose, the experimental fettuccine formulated with carob contained higher ($p < 0.05$) sucrose content than those formulations without this legume (e.g., P-60.0 vs. P-60.10 or PC-60.0 vs. PC-60.10).

Raffinose, ciceritol, and stachyose were detected in all of the uncooked and cooked fettuccine samples, except in the formulation elaborated with 100% rice and the commercial sample. The main α-galactoside detected in the uncooked and cooked samples was stachyose, accounting for around 55–84% of the total α-galactosides, except in P-20.0 and P-20.10 (samples with a higher content of raffinose). The total α-galactosides content was significantly higher in both the uncooked and the cooked samples formulated with carob bean fruit.

Table 3. Content of soluble sugars, ciceritol, and α-galactosides (mg/g dry weight) in raw materials and in the uncooked (P-) and cooked (PC-) fettuccine, as well as in the commercial rice pasta.

Sample	Sucrose	Maltose	Raffinose	Ciceritol	Stachyose	Total α-Galactosides
Bean	30.00 ± 0.95	n.d.	5.92 ± 0.09	0.34 ± 0.01	26.85 ± 0.25	32.77
Carob fruit	150.46 ± 10.04	n.d.	5.84 ± 0.02	n.d.	n.d.	5.84
Rice	2.98 ± 0.15	n.d.	n.d.	n.d.	n.d.	n.d.
P-20.0	8.59 ± 0.39 [b, A]	n.d.	6.40 ± 0.58 [e, A]	2.73 ± 0.30 [b, A]	3.95 ± 0.38 [a, A]	10.35 [d, A]
P-20.10	25.46 ± 0.09 [l, A]	n.d.	8.31 ± 0.16 [g, A]	2.19 ± 0.22 [b, A]	3.47 ± 0.25 [a, A]	11.78 [e, A]
P-40.0	14.18 ± 0.34 [f, A]	n.d.	9.11 ± 0.27 [h, A]	4.71 ± 0.08 [e, A]	9.06 ± 0.18 [b, A]	18.17 [h, A]
P-40.10	36.70 ± 0.78 [p, A]	n.d.	8.00 ± 0.80 [g, A]	6.43 ± 0.14 [f, A]	13.37 ± 0.32 [d, A]	21.37 [j, A]
P-60.0	23.77 ± 0.74 [k, A]	n.d.	9.05 ± 0.29 [h, A]	7.88 ± 0.17 [g, A]	17.25 ± 0.37 [f, A]	26.30 [k, A]
P-60.10	41.98 ± 0.68 [q, A]	n.d.	13.13 ± 0.22 [j, A]	10.09 ± 0.27 [h, A]	17.41 ± 0.36 [f, A]	30.54 [m, A]
P-80.0	29.63 ± 1.06 [n, A]	n.d.	11.84 ± 0.89 [i, A]	12.90 ± 0.94 [j, A]	26.92 ± 1.45 [i, A]	38.76 [n, A]
P-80.10	57.44 ± 0.52 [r, A]	n.d.	18.72 ± 0.20 [l, A]	13.55 ± 0.48 [k, A]	26.12 ± 0.41 [h, A]	44.85 [o, A]
P-Bean	36.53 ± 0.93 [p, A]	n.d.	13.82 ± 0.25 [k, A]	12.36 ± 0.18 [i, A]	30.71 ± 0.34 [j, A]	44.53 [o, A]
P-Rice	64.46 ± 1.44 [s, A]	64.77 ± 1.30 [d, A]	n.d.	n.d.	n.d.	n.d.
P-Commercial rice	18.32 ± 0.35 [h, A]	15.51 ± 1.21 [b, A]	n.d.	n.d.	n.d.	n.d.
PC-20.0	5.76 ± 0.44 [a, B]	n.d.	0.77 ± 0.08 [a, B]	1.37 ± 0.17 [a, B]	3.73 ± 0.06 [a, A]	4.51 [b, B]
PC-20.10	18.00 ± 0.28 [g, B]	n.d.	1.77 ± 0.09 [b, B]	2.54 ± 0.17 [b, A]	3.89 ± 0.12 [a, A]	5.67 [c, B]
PC-40.0	9.66 ± 0.35 [c, B]	n.d.	1.86 ± 0.23 [b, B]	4.06 ± 0.28 [d, B]	9.78 ± 0.18 [b, A]	11.64 [e, B]
PC-40.10	22.37 ± 0.54 [j, B]	n.d.	4.03 ± 0.31 [b, B]	4.82 ± 0.79 [e, B]	11.17 ± 1.02 [c, B]	15.20 [g, B]
PC-60.0	11.76 ± 0.16 [d, B]	n.d.	2.14 ± 0.11 [c, B]	3.41 ± 0.14 [c, B]	11.01 ± 0.09 [c, B]	13.15 [f, B]
PC-60.10	27.20 ± 1.81 [m, B]	n.d.	5.13 ± 0.23 [d, B]	3.52 ± 0.23 [c, B]	10.74 ± 0.50 [c, A]	15.87 [g, B]
PC-80.0	21.02 ± 0.55 [i, B]	n.d.	7.03 ± 0.05 [f, B]	7.66 ± 0.03 [g, B]	20.51 ± 0.72 [g, B]	27.54 [l, B]
PC-80.10	30.28 ± 0.81 [n, B]	n.d.	7.33 ± 0.12 [f, B]	4.70 ± 0.20 [e, B]	19.34 ± 0.23 [e, B]	26.67 [l, B]
PC-Bean	17.57 ± 0.41 [g, B]	n.d.	3.72 ± 0.29 [c, B]	3.31 ± 0.12 [c, B]	15.89 ± 0.35 [e, B]	19.61 [i, B]
PC-Rice	12.80 ± 0.11 [e, B]	10.85 ± 0.71 [a, B]	n.d.	n.d.	n.d.	n.d.
PC-Commercial rice	34.55 ± 0.33 [o, B]	24.06 ± 0.79 [c, B]	n.d.	n.d.	n.d.	n.d.

Values are means ± standard deviation ($n = 4$). Mean values in the same column followed by a different superscript letter are significantly different ($p < 0.05$); small superscript letters mean differences among all of the samples analyzed, whereas capital superscript letters mean differences due to the cooking treatment for the same formulation. n.d., not detected.

In general, the cooking process reduced ($p < 0.05$) the content of the individual sugars determined in each fettuccine sample. During the cooking process, pasta is hydrated and a softening of the food matrix occurs, facilitating the leaching of the sugars into the broth. The different sugars are not affected to the same extent by the cooking process, with the reduction of the individual sugars analyzed being from 4% (stachyose) to 88% (raffinose). This fact may be related to the different water solubility of these low-molecular weight sugars that migrate into the cooking water, which is discarded. These results are in accordance to those published by Laleg et al. [36] in gluten-free pasta elaborated exclusively from faba, lentil, or black-gram flours. The higher percentage of bean in the uncooked and cooked samples caused a higher α-galactosides content. Consequently, a good linear adjustment ($R^2 = 0.85$) to the model was observed: total α-galactosides (mg/g) = 0.858 + 0.359 × %bean. The sugar profile of the uncooked and cooked fettuccine elaborated with 100% rice and the commercial rice pasta did not contain α-galactosides—they only presented sucrose and maltose (50% of each sugar, approximately).

It should be noted that α-galactosides are traditionally associated with the production of flatulence, diarrhoea, and abdominal pain, since humans cannot digest these compounds, and they are used as substrates for anaerobic fermentation in the colon. Nowadays, α-galactosides are considered as nutritionally active factors or bioactive components with prebiotic activity by promoting the beneficial activity of intestinal microflora (bifidobacterias, bacteroides, and eubacteria), and they positively affect the immune system; thus, in humans they exert their health benefits through improving gut health, reducing constipation and diarrohea, stimulating the immune system, and increasing resistance to infection [16,37,38].

The commercial rice and the experimental fettuccine composed of 100% rice flour did not contain α-galactosides; thus, these samples did not have the health benefits associated with α-galactosides. The rest of the formulations showed high contents of α-galactosides and, as a consequence, they can exert health benefits associated with their consumption. There is not a dietary reference intake, although Martinez-Villaluenga et al. [39] reported that to obtain health benefits, an effective daily dose of α-galactosides is 3 g/day, since higher doses could be associated with the anti-nutritional problems mentioned above. Considering one serving of 60 g (dry weight) per day of fettuccine, a supply of between 0.67 g (PC-20.0) and 1.65 g (PC-80.0) of α-galactosides can be achieved. Thus, according to the above data [37], all of the experimental samples could promote health functions related to these compounds.

3.3. Trypsin and Chymotrypsin Inhibitors

Table 4 shows the trypsin and chymotrypsin inhibitor activities corresponding to the uncooked (P-) and cooked (PC-) fettuccine, as well as to the commercial rice pasta. The fettuccine (uncooked and cooked) elaborated with 100% rice and the commercial rice sample showed lower trypsin and chymotrypsin inhibitor activities, whereas P-Bean (13.19 TIU/mg and 16.87 CIU/mg) and PC-Bean (12.12 TIU/mg and 16.62 CIU/mg) were the samples with the highest protease inhibitor activities. The higher the bean content, the higher the inhibitor activities in all of the formulations; in fact, a strong linear correlation between the percentage of bean and protease inhibitor activities was fitted according to the models: TIU/mg = 0.454 + 0.122 × %bean ($R^2 = 0.98$) and CIU/mg = 0.597 + 0.156 × %bean ($R^2 = 0.92$).

Thermal processing of legumes can be applied to reduce totally or partially the protease inhibitor activities due to the fact that they are thermo-labile compounds [17,39]. In general, the cooked fettuccine did not show significant ($p > 0.05$) changes in these activities compared to the uncooked samples. A slight increase ($p > 0.05$) was observed in both types of protease inhibitor activities after cooking, except in some samples: PC-80.0 and PC-Bean for trypsin inhibitor activity, and PC-20.10, PC-40.0, PC-60.0, and PC-60.10 for chymotrypsin inhibitor activity. This increase in their content (on a dry weight basis) would be due to the cooking losses, as has been reported above for other compounds [40,41].

The slight changes observed after cooking could also be due to the short cooking time applied (on average 3 min), and to the different heat sensitivity of both types of protease inhibitors.

The results obtained in the experimental fettuccine are in concordance with the results of the trypsin inhibitor activity obtained by Zhao et al. [42] for uncooked and cooked wheat semolina spaghetti fortified with 20% of chickpea. On the contrary, Frias et al. [43] reported a reduction of trypsin inhibitor activity by around 52–59% after boiling macaroni elaborated with 100% green pea or yellow pea flours for 30 min. Even though pulse protease inhibitors interfere with protein digestibility [16], they are considered key players in several biological processes, including cancer progression, and they are associated with the capacity to prevent certain tumoral pathologies [5,42]. Thus, diets containing these compounds have been associated with a reduction in carcinogen-induced effects [41].

Table 4. Content of trypsin inhibitors (TIU/mg dry weight), chymotrypsin inhibitors (CIU/mg dry weight), and lectin content (%PHA) in the uncooked (P-) and cooked (PC-) fettuccine samples and in the commercial rice pasta.

Sample	Trypsin Inhibitors	Chymotrypsin Inhibitors	Lectin
Bean	23.21 ± 0.66	7.74 ± 0.28	0.59 ± 0.01
Carob fruit	0.30 ± 0.02	n.d.	-
Rice	0.15 ± 0.01	n.d.	-
P-20.0	2.33 ± 0.07 [b, A]	2.98 ± 0.25 [b c, A]	0.26 ± 0.04 [a]
P-20.10	3.10 ± 0.28 [c, A]	2.76 ± 0.23 [b, A]	0.36 ± 0.17 [a b]
P-40.0	5.67 ± 0.08 [d, A]	5.85 ± 0.54 [d, A]	0.38 ± 0.10 [a b]
P-40.10	6.19 ± 0.40 [d e, A]	8.22 ± 0.84 [g, A]	0.42 ± 0.12 [a b]
P-60.0	7.57 ± 0.56 [f, A]	10.85 ± 1.05 [h, A]	0.34 ± 0.18 [a b]
P-60.10	7.25 ± 0.51 [f, A]	10.23 ± 1.08 [h, A]	0.39 ± 0.12 [a b]
P-80.0	10.26 ± 0.77 [g, A]	16.15 ± 0.75 [j, A]	0.54 ± 0.10 [b]
P-80.10	10.19 ± 0.42 [g, A]	7.70 ± 0.32 [e f, A]	0.56 ± 0.28 [b]
P-Bean	13.19 ± 0.75 [j, A]	16.87 ± 0.57 [j, A]	0.58 ± 0.14 [b]
P-Rice	0.50 ± 0.04 [a, A]	0.56 ± 0.13 [a, A]	n.d.
P-Commercial rice	0.66 ± 0.03 [a, A]	0.40 ± 0.10 [a, A]	n.d.
PC-20.0	2.45 ± 0.07 [b, A]	3.41 ± 0.34 [b c, A]	n.d.
PC-20.10	3.17 ± 0.22 [c, A]	4.01 ± 0.86 [c, B]	n.d.
PC-40.0	5.67 ± 0.35 [d, A]	6.85 ± 0.61 [e, B]	n.d.
PC-40.10	6.35 ± 0.36 [e, A]	7.45 ± 0.74 [e f, A]	n.d.
PC-60.0	6.96 ± 0.37 [f, A]	12.26 ± 1.01 [i, B]	n.d.
PC-60.10	7.13 ± 0.72 [f, A]	12.08 ± 1.55 [i, B]	n.d.
PC-80.0	11.10 ± 0.34 [h, B]	16.30 ± 1.27 [j, A]	n.d.
PC-80.10	10.24 ± 0.74 [g, A]	8.21 ± 0.72 [g, A]	n.d.
PC-Bean	12.12 ± 0.35 [i, B]	16.62 ± 1.10 [j, A]	n.d.
PC-Rice	0.15 ± 0.02 [a, A]	0.47 ± 0.01 [a, A]	n.d.
PC-Commercial rice	0.33 ± 0.01 [a, A]	0.29 ± 0.02 [a, A]	n.d.

Values are means ± standard deviation ($n = 4$). Mean values in the same column followed by a different superscript letter are significantly different ($p < 0.05$); small superscript letters mean differences among all of the samples analyzed, whereas capital superscript letters mean differences due to the cooking treatment for the same formulation. n.d., not detected.

3.4. Lectin Content

Raw bean var. Almonga had a very low amount of lectin (Table 4), lower than the Pinto bean cultivar used as a negative (non-toxic) control; thus, their inclusion in foods that need a low cooking time can be of great interest to avoid the negative effect associated with raw or slightly cooked beans. The content of lectins in the different uncooked and cooked fettuccine evaluated by indirect ELISA is shown in Table 4. The results of ELISA show that the P-Bean sample presented the highest value of PHA (0.58%), while in the P-Rice and in P-Commercial rice samples, PHA was not detected. The uncooked fettuccine showed higher PHA values as the bean percentage in the samples increased, ranging from 0.26% to 0.58% (P-20.0 and P-Bean, respectively).

The addition of carob bean to the formulations did not significantly increase the PHA content (e.g., P-40.0 vs. P-40.10). After the cooking process, there was a total reduction of the lectin content in all samples analyzed, in agreement with previous works [11,17] that indicated that the heat processing of different formulations containing bean or lentil was effective to eliminate their lectins. Taking into account that PHA is considered a toxic lectin, their absence after cooking made the experimental fettuccine a safe food, suitable for consumption without the risks to health (vomiting, diarrhoea, or bloating) that appear after consumption of raw or uncooked samples with high lectin content.

3.5. Phenolic Composition and Antioxidant Activity

Table 5 presents the content of anthocyanins, flavonols, tartaric acid, and total phenols and their antioxidant activity (ORAC) determined in the uncooked (P-) and cooked (PC-) fettuccine and commercial samples. As it has been observed above for other bioactive compounds, the increase in the amount of bean in the formulation produced a rise in the total phenols content. Moreover, it is important to note that the supplementation of the different rice/bean samples with 10% WCF increases ($p < 0.05$) the content of phenolic compounds, both in the uncooked and the cooked samples.

The uncooked fettuccine (P-) showed, in general, higher total phenols content ($p < 0.05$) than the cooked samples (PC-), except the fettuccine elaborated with 100% rice that did not show significant differences after cooking. These results are in concordance with the data of the raw materials (Table 5), since WCF and bean were the source of polyphenols in the elaborated fettuccine. Regarding the present study, the losses of the total phenols content after cooking the fettuccine (around 17–48%) were higher than that reported by Pedrosa et al. [16] after the canning process of Almonga bean seeds (by about 11%), but were similar to the results found in the literature in cooked pasta fortified with buckwheat, common and black beans, and red and black lentil pasta [18,43,44]. According to the literature [18,43–45], the reduction in phenols can be attributed to the synergistic combination of several factors, such as the pasta-making process, the sensitivity of heat during the cooking process, the leaching into the broth because of the softening of the food matrix that occurs during pasta cooking, as well as the association with the food matrix during cooking, that might hinder their extraction. Verardo et al. [46] reported for cooked buckwheat spaghetti that 11.6% of the different phenolic compounds were dissolved in the cooking water, and the pasta-making process caused a loss of 45.9% of the total phenolic compounds present in the raw materials used in the elaboration of buckwheat spaghetti.

Compared to the cooked commercial rice pasta, the cooked fettuccine fortified with WCF (10%) contained from 2.5-fold (PC-20.10) to 3.2-fold (PC-80.10) more phenols than the commercial sample, similar to other durum wheat pasta enriched with 1–5% carob pods [12].

Tartaric ester, flavonol, and anthocyanin content of the experimental fettuccine was also affected by the cooking process; the tartaric esters were reduced by around 7–50% and the flavonols between 10% and 75%, while the commercial rice sample did not reduce significantly their content after the cooking process. The anthocyanins were reduced ($p < 0.05$) in all the samples by the cooking process. The reduction ranged from 52% to 82% in the fettuccine without WCF (PC-20.0 and PC-60.0) and from 68% to 86% in the fettuccine with WCF (PC-20.10 and PC-60.10). The decrease was small in the samples with an 80% bean amount, where a reduction of 7% was observed; although the cause of this small reduction remains uncertain, a possible hypothesis is that in these samples, during the pasta-making and cooking processes, a higher amount of phenols are able to bind to the food matrix, which prevents its extraction [44].

In relation to the antioxidant activity (ORAC assay) (Table 5), the results show that the highest value corresponded to the uncooked sample P-80.10 (13.68 µmol Trolox/g), and the lowest corresponded to the P-Commercial Rice (4.38 µmol Trolox/g). The antioxidant activity was linked to the presence of phenols supplied by the bean and WCF flours, and to the presence of other antioxidant compounds such as the Maillard reaction products formed during the pasta-making drying step. As mentioned above for the phenolic compounds, the presence of WCF increased the antioxidant activity in comparison to the same formulations without this legume. After cooking, a slight drop ($p > 0.05$) of the antioxidant activity was observed; the ORAC values ranged from 3.08 to 10.32 µmol Trolox/g for the PC-Commercial rice and PC-60.10 samples, respectively. A similar decrease was reported by Rocchetti et al. [47] in six commercial GF pastas elaborated with black rice, chickpea, red lentil, sorghum, amaranth, and quinoa. The correlation between total phenols, tartaric esters, flavonols, and ORAC values was high, with a Pearson's coefficient R^2 of 0.88, 0.84, and 0.73, respectively (Table S1).

Table 5. Anthocyanins (µg C3GE/g dry weight), flavonols (µg QE/g dry weight), tartaric esters (mg CAE/g dry weight) and total phenols (mg (+) CE/g dry weight) content and the antioxidant activity (µmol Trolox/g dry weight) of the raw materials and the uncooked (P-) and cooked (PC-) fettuccine, as well as the commercial rice pasta.

Sample	Total Phenols	Tartaric Esters	Flavonols	Anthocyanins	Antioxidant Activity (ORAC)
Bean	2.88 ± 0.02	0.21 ± 0.01	0.08 ± 0.001	36.96 ± 0.24	24.33 ± 0.07
Carob fruit	20.73 ± 0.10	0.72 ± 0.01	0.75 ± 0.001	18.00 ± 0.15	69.89 ± 1.62
Rice	0.90 ± 0.03	0.02 ± 0.001	0.03 ± 0.001	18.70 ± 0.83	3.80 ± 0.30
P-20.0	2.88 ± 0.20 [d e, A]	0.07 ± 0.001 [c d, A]	0.03 ± 0.001 [c, A]	10.65 ± 0.001 [g h, A]	5.26 ± 0.65 [c d, A]
P-20.10	5.77 ± 0.06 [l, A]	0.14 ± 0.02 [i, A]	0.10 ± 0.001 [g, A]	10.64 ± 0.001 [g h, A]	10.49 ± 0.69 [g h, A]
P-40.0	4.58 ± 0.01 [g h, A]	0.09 ± 0.001 [e f, A]	0.03 ± 0.001 [b c, A]	8.25 ± 0.001 [f g h, A]	6.94 ± 0.67 [f, A]
P-40.10	5.40 ± 0.21 [j k, A]	0.14 ± 0.001 [i, A]	0.08 ± 0.001 [f, A]	9.16 ± 0.001 [f g h, A]	9.67 ± 1.23 [h i, A]
P-60.0	4.70 ± 0.40 [g h i, A]	0.09 ± 0.001 [f, A]	0.03 ± 0.001 [c, A]	8.25 ± 0.001 [d e f g h, A]	9.20 ± 0.52 [g h, A]
P-60.10	5.80 ± 0.31 [k l, A]	0.15 ± 0.001 [j, A]	0.10 ± 0.001 [g, A]	6.32 ± 0.001 [b c d e f, A]	12.39 ± 0.27 [k, A]
P-80.0	4.48 ± 0.34 [g, A]	0.12 ± 0.001 [g, A]	0.04 ± 0.001 [d, A]	5.70 ± 0.001 [b c d e f, A]	11.06 ± 0.64 [j, A]
P-80.10	7.27 ± 0.21 [m, A]	0.20 ± 0.01 [k, A]	0.10 ± 0.001 [g, A]	6.87 ± 0.001 [c d e f g, A]	13.68 ± 0.58 [l, A]
P-Bean	5.11 ± 0.13 [i j, A]	0.12 ± 0.001 [g h, A]	0.04 ± 0.001 [d, A]	7.05 ± 0.001 [c d e f g, A]	12.46 ± 0.19 [k, A]
P-Rice	2.28 ± 0.10 [a b c, A]	0.06 ± 0.001 [a b, A]	0.02 ± 0.001 [a b, A]	12.43 ± 0.001 [h, A]	4.57 ± 0.25 [b c, A]
P-Commercial rice	2.50 ± 0.12 [b c d, A]	0.08 ± 0.001 [e, A]	0.02 ± 0.001 [a b, A]	2.78 ± 0.001 [a b, A]	4.38 ± 0.42 [a b, A]
PC-20.0	2.35 ± 0.20 [b c, B]	0.09 ± 0.001 [e f, B]	0.02 ± 0.001 [a b, B]	1.77 ± 0.001 [a, B]	5.12 ± 0.54 [c d, A]
PC-20.10	4.47 ± 0.07 [g, B]	0.13 ± 0.001 [h i, A]	0.09 ± 0.001 [f, B]	3.09 ± 0.001 [a b c, B]	8.20 ± 0.74 [g h, A]
PC-40.0	3.52 ± 0.22 [f, B]	0.07 ± 0.001 [c d, B]	0.02 ± 0.001 [a b, B]	3.89 ± 0.01 [a b c, B]	5.97 ± 0.36 [f, A]
PC-40.10	4.59 ± 0.25 [g h, B]	0.13 ± 0.001 [h i, A]	0.08 ± 0.001 [f, A]	2.91 ± 0.01 [a b c, B]	10.01 ± 0.62 [h i, A]
PC-60.0	3.21 ± 0.27 [e f, B]	0.06 ± 0.001 [c d, B]	0.02 ± 0.001 [a b, B]	1.37 ± 0.001 [a, B]	8.77 ± 0.86 [g h, A]
PC-60.10	5.03 ± 0.26 [h i j, B]	0.14 ± 0.001 [j, A]	0.10 ± 0.001 [g, A]	0.91 ± 0.001 [a, B]	10.32 ± 1.22 [k, A]
PC-80.0	3.29 ± 0.20 [e f, B]	0.09 ± 0.001 [f, B]	0.03 ± 0.001 [c, B]	5.33 ± 0.001 [a b c d e, A]	8.09 ± 0.72 [j, A]
PC-80.10	5.83 ± 0.55 [k l, B]	0.14 ± 0.001 [i, B]	0.07 ± 0.001 [e, B]	6.41 ± 0.001 [b c d e f, A]	9.31 ± 0.60 [l, A]
PC-Bean	2.66 ± 0.18 [c d, B]	0.06 ± 0.001 [b c, B]	0.01 ± 0.001 [a, B]	3.22 ± 0.001 [a b c, A]	6.99 ± 0.65 [k, A]
PC-Rice	2.11 ± 0.06 [a b, A]	0.05 ± 0.001 [a, A]	0.02 ± 0.001 [a b, A]	13.19 ± 0.001 [i, B]	4.66 ± 0.74 [b c, A]
PC-Commercial rice	1.80 ± 0.07 [a, B]	0.07 ± 0.001 [d, A]	0.01 ± 0.001 [a, A]	2.63 ± 0.001 [a b c d, A]	3.08 ± 0.51 [a b, A]

Values are means ± standard deviation ($n = 4$). Mean values in the same column followed by a different superscript letter are significantly different ($p < 0.05$); small superscript letters mean differences among all of the samples analyzed, whereas capital superscript letters mean differences due to the cooking treatment for the same formulation. C3GE, cyanidin-3-glucoside equivalents; QE, quercetin equivalents; CAE, caffeic acid equivalents; CE, (+) catechin equivalents.

It is well known that phenols have antioxidant, anti-inflammatory, and antimicrobial properties [48]. Consequently, they are preventive agents against several degenerative diseases, and in vitro, they are associated with a low incidence of several types of cancer (breast, skin, prostate, and colon cancer), as well as cardiovascular diseases [41]. Carob flour was described by Avallone et al. [14] as a rich source of phenolic compounds linked to antioxidant and cytotoxic activities. The daily phenols intake depends on specific dietary preferences and socio-cultural factors of different countries. In general, it is accepted that the total polyphenol intake for the overall population is about 900 mg/day [48,49]. At present, there is no a dietary recommended intake of phenolic compounds to exert these healthy effects; although, some authors have reported a minimum dose of 300 mg of total phenols to obtain health benefits. In this regard, one serving (60 g) of the experimental rice/bean fettuccine supplemented with 10% WCF could supply, on average, 300 mg of total phenols, that, in addition, represent a third of the daily intake of phenols. Thus, the consumption of the experimental rice/bean-based fettuccine supplemented with 10% WCF would be valuable for a balanced diet with some health-related functions.

A principal component analysis (PCA) was carried out on the results obtained regarding the different bioactive compounds and the antioxidant activity (ORAC) of the uncooked (P-) and cooked (PC-) experimental fettuccine and the commercial rice pasta (control). The first three principal components explained 84.29% of the total variance (supplementary PCA file, Figures S2 and S3). The first two principal components (Figure S2) explained a relative percentage of 72.56%. PC1 was most correlated with total IP, total α-galactosides, tartaric esters, and antioxidant activity, and PC2 was most correlated with protease inhibitors and flavonols. The percentage of legumes in the experimental fettuccine was well-described by PC1, and the cooking effect was well–described by both PC1 and

PC2. Lectin and anthocyanin content contributed, to a higher extent, to explaining the variance of PC3 (11.74%).

3.6. Color Analysis

Color is a primary parameter for consumers' acceptance of a product such as dry and cooked pasta. Food color without additives mainly depends on the composition of raw materials (protein, dietary fiber, phenols, etc.) used in the elaboration of the pasta [15]. In general, it was observed that an enrichment of wheat pasta samples with other kinds of flours caused an overall darkening of the pasta color shade, yielding a significant decrease in L* value [50].

The color values of the uncooked and cooked fettuccine, as well as those of the commercial sample, are shown in Table 6. The increase in bean percentage in the formulation did not cause significant changes in the luminosity (L*) of the uncooked fettuccine; however, there was an increase ($p < 0.05$) in the yellow hue (b*), similarly to that reported by Gallegos-Infante et al. [18]. The color parameters of both the cooked and uncooked fettuccine showed significant differences ($p < 0.05$) between the samples with or without WCF. The fettuccine supplemented with 10% WCF had lower values of L*, and they resulted in redder (a*) ($p < 0.05$) and less yellowish samples in comparison to the formulations with equal bean content but without WCF. These differences can be linked to the brown color of the raw WCF and the processing conditions, both in the making and the cooking processes [1,15].

Table 6. Effects of the cooking process on the color values in the CIELab space of the uncooked (P-) and cooked (PC-) fettuccine and the commercial rice pasta.

Formulation	Uncooked Pasta (P-)			Cooked Pasta (PC-)		
	L*	a*	b*	L*	a*	b*
20.0	84.56 ± 0.21 [g h, A]	1.42 ± 0.03 [f, A]	16.62 ± 0.14 [h, A]	65.03 ± 1.30 [d, B]	−1.57 ± 0.17 [a b, B]	12.37 ± 1.26 [c, B]
20.10	69.31 ± 0.15 [e, A]	4.58 ± 0.07 [j, A]	14.59 ± 0.17 [e f, A]	44.44 ± 1.42 [a, B]	7.37 ± 0.63 [m, B]	12.60 ± 0.78 [c d, B]
40.0	84.98 ± 0.12 [h, A]	1.72 ± 0.05 [f g, A]	18.16 ± 0.23 [i, A]	65.69 ± 0.61 [d, B]	−1.27 ± 0.21 [b c, B]	15.00 ± 0.76 [f, B]
40.10	70.07 ± 0.12 [e, A]	4.27 ± 0.06 [j, A]	15.39 ± 0.12 [f g, A]	45.73 ± 1.17 [a b, B]	6.42 ± 0.47 [l, B]	12.56 ± 0.42 [c d, B]
60.0	83.25 ± 0.28 [f g, A]	2.51 ± 0.06 [h, A]	20.56 ± 0.44 [j, A]	65.63 ± 1.08 [d, B]	−1.04 ± 0.30 [c, B]	16.23 ± 1.19 [g h, B]
60.10	69.05 ± 0.29 [e, A]	4.23 ± 0.01 [j, A]	15.41 ± 0.11 [f g, A]	47.06 ± 1.48 [b, B]	5.68 ± 0.50 [k, B]	11.58 ± 0.63 [b c, B]
80.0	83.66 ± 0.12 [g h, A]	2.12 ± 0.03 [g h, A]	19.58 ± 0.08 [i, A]	65.43 ± 1.19 [d, B]	−0.50 ± 0.4 [d, B]	17.77 ± 1.24 [i, B]
80.10	68.44 ± 0.27 [e, A]	4.69 ± 0.08 [j, A]	18.29 ± 0.27 [i, A]	46.26 ± 1.04 [b, B]	5.84 ± 0.52 [k, B]	13.63 ± 0.63 [d e, B]
Bean 100%	81.93 ± 0.16 [f, A]	3.44 ± 0.03 [i, A]	22.85 ± 0.14 [k, A]	64.53 ± 1.20 [d, B]	0.63 ± 0.35 [e, B]	19.98 ± 0.61 [j, B]
Rice 100%	87.02 ± 0.10 [i, A]	0.69 ± 0.01 [e, A]	11.55 ± 0.05 [b c, A]	62.93 ± 2.09 [c, B]	−1.86 ± 0.11 [a, B]	4.42 ± 0.61 [a, B]
Commercial rice	92.01 ± 0.05 [j, A]	0.30 ± 0.001 [e, A]	11.07 ± 0.10 [b, A]	70.08 ± 1.36 [e, B]	−1.60 ± 0.19 [a b, B]	12.66 ± 1.54 [c d, B]

Values are means ± standard deviation ($n = 12$). Mean values in the same column followed by a different superscript letter are significantly different ($p < 0.05$); small superscript letters mean differences among all of the samples analyzed, whereas capital superscript letters mean differences due to the cooking treatment for the same formulation. L*, luminosity; a*, redness/greenness; b*, yellowness/blueness.

These changes in the luminosity were strongly and positively correlated to the fortification with 10% WCF: L* = 69.563 − 1.902 × %WCF ($R^2 = -0.72$); which is in concordance with the fortified pasta elaborated with 5% carob fiber reported by Chillo et al. [1]. The cooked fettuccine without WCF resulted in a less-red and less-yellow pasta than the cooked samples with WCF. In general, the cooked pasta was darker with brown hues, probably due to the transformation of carbohydrates by the reactions of Maillard during thermal processing [43]. The dry (P-) and cooked (PC-) commercial pasta had the highest lightness values (92.01 and 70.08, respectively) of all of the analyzed samples, while the fettuccine elaborated with 100% bean were the most yellow ones (22.85 and 19.98, P-Bean and PC-Bean, respectively). In the market, many types of colored pastas are available (whole grain, fortified with vegetables, inks, etc.), Thus, although the color of the novel pasta is different from the control pasta (Figure S1), they would be easily accepted by consumers.

3.7. Texture Analysis

Regarding the textural parameters of cooked pastas, firmness and stickiness play an important role in the acceptability of pasta by consumers, where a sticky pasta is generally unacceptable [46].

The texture values (Table 7) of the GF pasta elaborated with rice and bean show an increase in the hardness as the percentage of bean increased in the formulation, corresponding to the highest hardness of 70.36 N in the PC-Bean fettuccine, as per the occurrence in precooked rice–yellow pea pasta and faba bean pasta reported by Bouasla et al. and Rosa-Sibakov et al. [50,51], respectively. Nevertheless, the supplementation of fettuccine with 10% WCF decreased the hardness compared to the same pasta without WCF, probably due to the increase in dietary fiber content, which may have led to the formation of crashes or breaks inside the fettuccine strand, thus weakening the pasta structure [46,52]. The adhesiveness of fettuccine, in general, showed a slight increase with the increase in the bean percentage; on the contrary, a higher percentage of bean produced lower stickiness. Bouasla et al. [51] reported the same tendency in precooked rice–yellow pea pasta.

Table 7. Instrumental measure of the texture of the cooked fettuccine (PC-) and the commercial pasta (control).

Sample	Hardness (N)	Stickiness (N)	Adhesiveness (N·s)	Cutting Firmness (N)	Cutting Consistency (N·s)
PC-20.0	58.31 ± 0.62 [d]	9.21 ± 0.19 [a]	−0.44 ± 0.12 [a]	3.14 ± 0.04 [d e]	5.39 ± 0.06 [b]
PC-20.10	44.49 ± 0.44 [b]	6.66 ± 0.16 [b c]	−0.41 ± 0.09 [a b c]	3.14 ± 0.02 [d e]	5.39 ± 0.04 [b]
PC-40.0	55.56 ± 0.21 [c d]	6.96 ± 0.20 [b c]	−0.39 ± 0.11 [a b c]	3.23 ± 0.02 [d e]	6.17 ± 0.05 [c d]
PC-40.10	53.51 ± 0.25 [c d]	5.88 ± 0.18 [c d]	−0.35 ± 0.07 [b c d]	4.02 ± 0.08 [f]	5.49± 0.10 [b c]
PC-60.0	58.21 ± 0.37 [d]	4.61 ± 0.12 [d]	−0.30 ± 0.06 [d]	3.33 ± 0.04 [e]	6.47 ± 0.08 [d]
PC-60.10	53.21 ± 0.42 [c]	6.47 ± 0.12 [c]	−0.40 ± 0.05 [a b c]	2.74 ± 0.04 [b c]	5.29 ± 0.09 [b]
PC-80.0	63.50 ± 0.46 [e]	7.94 ± 0.14 [a b]	−0.45 ± 0.07 [a]	4.02 ± 0.03 [f]	8.13 ± 0.09 [e]
PC-80.10	68.21± 0.70 [f]	6.17 ± 0.13 [c]	−0.42 ± 0.06 [a b]	2.94 ± 0.07 [c d]	6.66± 0.09 [d]
PC-Bean	70.36 ± 0.65 [f]	4.51 ± 0.15 [d]	−0.32 ± 0.09 [c d]	4.70 ± 0.03 [g]	10.58 ± 0.15 [f]
PC-Rice	35.28 ± 1.17 [a]	9.11 ± 0.35 [a]	−0.39 ± 0.10 [a b]	1.96 ± 0.04 [a]	3.53 ± 0.08 [a]
PC-Commercial	31.26 ± 0.32 [a]	2.74 ± 0.04 [e]	−0.37 ± 0.09 [a b c d]	2.45 ± 0.03 [b]	8.53± 0.11 [e]

Values are means ± standard deviation ($n = 12$). Mean values in the same column followed by a different superscript letter are significantly different ($p < 0.05$).

The addition of 10% WCF in the formulations produced, in general, a decrease in both the adhesiveness and the stickiness, although the changes in the adhesiveness were non-significant. The cooked samples with higher stickiness corresponded to the PC-20.0 and PC-Rice samples (9.21 N and 9.11 N, respectively). A reduction of the stickiness was also reported by other authors in amaranthus whole meal flour-based pasta fortified with quinoa, broad bean, or chickpea flour [1]. Both parameters are related to the amount of starch granules in the surface of the cooked pasta, the amount and type of dietary fiber, and the protein content. The manufacturing process utilized can also affect the textural parameters [2]. During the elaboration of the pasta by cold extrusion, the screw produced a mechanical heat able to pre-gelatinize some amount of the rice starch [53]. This pre-gelatinized starch, together with the unmodified starch granules, the high amount of fiber and protein, as well as the presence of caroubin (with similar properties of gluten), is responsible of the reduction in stickiness after cooking.

Considering the values obtained regarding the firmness and consistency from the cutting analysis, it can be observed that an increase in the bean content produced an increase in the value of both parameters. These increases were significant for samples with more than a 60% bean content, with the corresponding maximum values belonging to the PC-Bean sample (4.7 N and 10.58 N·s, respectively), while the minimum values corresponded to the fettuccine elaborated with 100% rice (1.96 N and 3.53 N·s, respectively). The addition of WCF produced a decrease in both parameters, related to the weakening of the fettuccine matrix due to the increase in the dietary fiber and the decrease of starch content of the fettuccine [15], which is in concordance with the hardness values obtained above.

In general, the elaborated fettuccine showed higher values of hardness (1.1–2 times) and stickiness (1.6–3.3 times), similar adherence, higher values of firmness (1.1–1.9 times), and lower consistency values (1.1–2.5 times) than the commercial sample analyzed. A firm and hard texture would be desirable because it allows a correct hydration of the pasta, which restricts the swelling of starch granules and prevents the easy fracturability that characterizes gluten-free pasta [36,54].

The changes in the texture parameters could also be related to the presence of bioactive compounds in the fettuccine. Taherian et al. [55] and Tsai et al. [56] reported that phytic acid (IP6) and phenols, respectively, can induce changes in the hardness of food products. To estimate the possible influence of bioactive compounds on the texture parameters, correlation Pearson's coefficients were calculated. The results of the hardness and firmness parameters correlated well with most of the bioactive compounds analyzed (Table S2). In the case of hardness, a strong positive correlation was observed with total inositol phosphates ($R^2 = 0.84$; $p = 0.0000$), total galactosides ($R^2 = 0.73$; $p = 0.0000$), and protease inhibitors ($R^2 = 0.81$; $p = 0.0000$). The firmness showed a moderate and positive correlation with the same bioactive compounds, with Pearson's coefficients of 0.58 ($p = 0.0000$), 0.57 ($p = 0.0000$), and 0.63 ($p = 0.0000$) for inositol phosphates, galactosides, and protease inhibitors, respectively.

4. Conclusions

The experimental fettuccine revealed a great health potential due to their bioactive compound composition. The inositol phosphates and protease inhibitors showed a slight increase, whereas α-galactosides, total phenols content, and antioxidant activity showed a decrease after cooking. Even though there is not a recommended dietary intake, according to the data found in the literature, the amount detected of these bioactive compounds in the different fettuccine would be enough to maintain their healthy characteristics (e.g., prebiotics, antioxidants, anticarcinogenic, etc.) and to reduce their associated drawbacks (e.g., flatulence, reduction of mineral bioavailability, impairment of digestion, etc.). The lectin content was eliminated by the cooking process, avoiding the toxic problems associated with their consumption. The addition of high percentages of bean improved the bioactive compound content, as well as the texture of the fettuccine. The pasta supplemented with 10% whole carob fruit was darker with brown hues, and showed better texture parameters (less hard, less sticky, and less adhesive) than the formulations without this legume. The improvement of the texture parameters observed could meet the expectation of consumers of GF products.

In comparison to the commercial sample, except that of 100% rice, all of the experimental fettuccine contained higher amounts of bioactive compounds, notably those supplemented with carob fruit, being a suitable nutritional and healthy alternative to the commercially available gluten-free rice-based pasta for celiac individuals and for the general population that consume gluten-free products.

Supplementary Materials: The following are available online at http://www.mdpi.com/2304-8158/9/4/415/s1. Table S1: Correlation coefficients between the phenolic compounds and the antioxidant activity; Table S2: Correlation coefficients between the bioactive compounds and the instrumental texture parameters of cooked fettuccine; Figure S1: The uncooked experimental rice/bean fettuccine without (20.0, 40.0, 60.0, 80.0, bean 100%, rice 100%) and with (20.10, 40.10, 60.10, and 80.10) whole carob fruit and the commercial rice pasta (control); Figure S2: Principal components analysis (PCA) projection of the PC1 and PC2 principal components. P-, uncooked pasta; PC-, cooked pasta. Parameters: Total inositol phosphates (IP Total), total α-galactosides, sucrose, protease inhibitors (TIU and CIU), lectins, total phenolics, tartaric esters, flavonols, anthocyanins, and antioxidant capacity (ORAC); Figure S3: Principal components analysis (PCA) projection of the PC1, PC2, and PC3 principal components. P-, uncooked pasta; PC-, cooked pasta. Parameters: Total inositol phosphates (IP Total), total α-galactosides, sucrose, protease inhibitors (TIU and CIU), lectins, total phenolics, tartaric esters, flavonols, anthocyanins, and antioxidant capacity (ORAC).

Author Contributions: M.M.P. conceived and designed the research experiments. C.A., B.C., C.C., E.G., and M.M.P. performed the experiments. C.A., B.C., C.C., E.G., and M.M.P. analyzed the data. C.A., B.C., C.C., E.G., and M.M.P. wrote, reviewed, and edited the paper. The manuscript was critically revised and approved by all authors. All authors have read and agreed to the published version of the manuscript.

Funding: The Spanish Ministry of Economy and Competitiveness (Project RTA2012-00042-C02 and Project RTA2015-00003-C02-01) supported this work. C.A. was supported by a contract (CPR2014-0068) from INIA, and co-financed by the European Regional Development Fund (FEDER) and the European Social Fund (FSE).

Conflicts of Interest: The authors declare no conflicts of interest.

References

1. Chillo, S.; Laverse, J.; Falcone, P.M.; Protopapa, A.; Del Nobile, M.A. Influence of the addition of buckwheat flour and durum wheat bran on spaghetti quality. *J. Cereal Sci.* **2008**, *47*, 144–152. [CrossRef]
2. Foschia, M.; Horstmann, S.W.; Arendt, E.K.; Zannini, E. Legumes as functional ingredients in gluten-free bakery and pasta products. *Annu. Rev. Food Sci. Technol.* **2017**, *8*, 75–96. [CrossRef] [PubMed]
3. Cabrera-Chávez, F.; de la Barca, A.M.C.; Islas-Rubio, A.R.; Marti, A.; Marengo, M.; Pagani, M.A.; Bonomi, F.; Iametti, S. Molecular rearrangements in extrusion processes for the production of amaranth-enriched, gluten-free rice pasta. *LWT* **2012**, *47*, 421–426. [CrossRef]
4. Duranti, M. Grain legume proteins and nutraceutical properties. *Fitoterapia* **2006**, *77*, 67–82. [CrossRef] [PubMed]
5. Arribas, C.; Cabellos, B.; Cuadrado, C.; Guillamón, E.; Pedrosa, M. Extrusion effect on proximate composition, starch and dietary fibre of ready-to-eat products based on rice fortified with carob fruit and bean. *LWT Food Sci. Technol.* **2019**, *111*, 387–393. [CrossRef]
6. Elliott, C. The nutritional quality of gluten-free products for children. *Pediatrics* **2018**, *142*, e20180525. [CrossRef] [PubMed]
7. Olmedilla-Alonso, B.; Pedrosa, M.M.; Cuadrado, C.; Brito, M.; Asensio-S-Manzanera, C.; Asensio-Vegas, C. Composition of two Spanish common dry beans (Phaseolus vulgaris), 'Almonga' and 'Curruquilla', and their postprandial effect in type 2 diabetics. *J. Sci.* **2013**, *93*, 1076–1082.
8. Giuberti, G.; Gallo, A.; Cerioli, C.; Fortunati, P.; Masoero, F. Cooking quality and starch digestibility of gluten free pasta using new bean flour. *Food Chem.* **2015**, *175*, 43–49. [CrossRef]
9. Turfani, V.; Narducci, V.; Durazzo, A.; Galli, V.; Carcea, M. Technological, nutritional and functional properties of wheat bread enriched with lentil or carob flours. *LWT* **2017**, *78*, 361–366. [CrossRef]
10. Goulas, V.; Stylos, E.; Chatziathanasiadou, M.V.; Mavromoustakos, T.; Tzakos, A.G. Functional components of carob fruit: Linking the chemical and biological space. *IJMS* **2016**, *17*, 1875. [CrossRef]
11. Arribas, C.; Cabellos, B.; Cuadrado, C.; Guillamón, E.; Pedrosa, M.M. Bioactive compounds, antioxidant activity, and sensory analysis of rice-based extruded snacks-like fortified with bean and carob fruit flours. *Foods* **2019**, *8*, 381. [CrossRef]
12. Biernacka, B.; Dziki, D.; Gawlik-Dziki, U.; Różyło, R.; Siastała, M. Physical, sensorial, and antioxidant properties of common wheat pasta enriched with carob fiber. *LWT* **2017**, *77*, 186–192. [CrossRef]
13. Feillet, P.; Roulland, T.M. Caroubin: A gluten-like protein isolated from carob bean germ. *Cereal Chem.* **1998**, *75*, 488–492. [CrossRef]
14. Avallone, R.; Plessi, M.; Baraldi, M.; Monzani, A. Determination of Chemical Composition of Carob (Ceratonia siliqua): Protein, Fat, Carbohydrates, and Tannins. *J. Food Compos. Anal.* **1997**, *10*, 166–172. [CrossRef]
15. Sęczyk, Ł.; Świeca, M.; Gawlik-Dziki, U. Effect of carob (*Ceratonia siliqua* L.) flour on the antioxidant potential, nutritional quality, and sensory characteristics of fortified durum wheat pasta. *Food Chem.* **2016**, *194*, 637–642. [CrossRef]
16. Muzquiz, M.; Varela, A.; Burbano, C.; Cuadrado, C.; Guillamón, E.; Pedrosa, M.M. Bioactive compounds in legumes: Pronutritive and antinutritive actions. Implications for nutrition and health. *Phytochem. Rev.* **2012**, *11*, 227–244. [CrossRef]
17. Pedrosa, M.M.; Cuadrado, C.; Burbano, C.; Muzquiz, M.; Cabellos, B.; Olmedilla-Alonso, B.; Asensio-Vegas, C. Effects of industrial canning on the proximate composition, bioactive compounds contents and nutritional profile of two Spanish common dry beans (*Phaseolus vulgaris* L.). *Food Chem.* **2015**, *166*, 68–75. [CrossRef]
18. Gallegos-Infante, J.-A.; Bello-Perez, L.A.; Rocha-Guzman, N.E.; Gonzalez-Laredo, R.F.; Avila-Ontiveros, M. Effect of the addition of common bean (*Phaseolus vulgaris* L.) flour on the in vitro digestibility of starch and undigestible carbohydrates in spaghetti. *J. Food Sci.* **2010**, *75*, H151–H156. [CrossRef]
19. Arribas, C.; Cabellos, B.; Guillamón, E.; Pedrosa, M.M. (FOODCHEM-D-20-02223). Cooking and sensory quality, nutritional composition and dietary fibre content of cold-extruded rice/bean fettucine fortified with whole carob fruit flour. *Food Chem.* (in press).
20. Burbano, C.; Muzquiz, M.; Ayet, G.; Cuadrado, C.; Pedrosa, M.M. Evaluation of antinutritional factors of selected varieties of Phaseolus vulgaris. *J. Sci.* **1999**, *79*, 1468–1472.

21. Pedrosa, M.M.; Cuadrado, C.; Burbano, C.; Allaf, K.; Haddad, J.; Gelencsér, E.; Takács, K.; Guillamón, E.; Muzquiz, M. Effect of instant controlled pressure drop on the oligosaccharides, inositol phosphates, trypsin inhibitors and lectins contents of different legumes. *Food Chem.* **2012**, *131*, 862–868. [CrossRef]

22. Welham, T.; Domoney, C. Temporal and spatial activity of a promoter from a pea enzyme inhibitor gene and its exploitation for seed quality improvement. *Plant. Sci.* **2000**, *159*, 289–299. [CrossRef]

23. Cuadrado, C.; Hajos, G.; Burbano, C.; Pedrosa, M.M.; Ayet, G.; Muzquiz, M.; Pusztai, A.; Gelencser, E. Effect of Natural Fermentation on the Lectin of Lentils Measured by Immunological Methods. *Food Agric. Immunol.* **2002**, *14*, 41–49. [CrossRef]

24. Dueñas, M.; Fernández, D.; Hernández, T.; Estrella, I.; Muñoz, R. Bioactive phenolic compounds of cowpeas (*Vigna sinensis* L). Modifications by fermentation with natural microflora and with Lactobacillus plantarum ATCC 14917. *J. Sci.* **2005**, *85*, 297–304.

25. Oomah, B.D.; Cardador-Martínez, A.; Loarca-Piña, G. Phenolics and antioxidative activities in common beans (*Phaseolus vulgaris* L). *J. Sci.* **2005**, *85*, 935–942.

26. Gallegos-Infante, J.; Rocha-Guzman, N.; Gonzalez-Laredo, R.; Ochoa-Martínez, L.; Corzo, N.; Bello-Perez, L.A.; Medina-Torres, L.; Peralta-Alvarez, L. Quality of spaghetti pasta containing Mexican common bean flour (*Phaseolus vulgaris* L.). *Food Chem.* **2010**, *119*, 1544–1549. [CrossRef]

27. AACC. Approved Methods of Analysis, 11th Ed. In *Method 66-50—Pasta and Noodle Cooking Quality-Firmness*; Cereals & Grains Association: St. Paul, MN, USA, 2000; pp. 1–3.

28. Roy, F.; Boye, J.I.; Simpson, B.K. Bioactive proteins and peptides in pulse crops: Pea, chickpea and lentil. *Food Res. Int.* **2010**, *43*, 432–442. [CrossRef]

29. Plaami, S.; Kumpulainen, J. Inositol phosphate content of some cereal-based foods. *J. Food Compos. Anal.* **1995**, *8*, 324–335. [CrossRef]

30. Anton, A.; Fulcher, R.S. Physical and nutritional impact of fortification of corn starch based extruded snacks with common bean (*Phaseolus vulgaris* L.) flour: Effects of bean addition and extrusion cooking. *Food Chem.* **2009**, *133*, 989–996. [CrossRef]

31. Sanz-Penella, J.M.; Wronkowska, M.; Soral-Smietana, M.; Haros, M. Effect of whole amaranth flour on bread properties and nutritive value. *LWT Food Sci. Technol.* **2013**, *50*, 679–685. [CrossRef]

32. Tazrart, K.; Zaidi, F.; Lamacchia, C.; Haros, M. Effect of durum wheat semolina substitution with broad bean flour (Vicia faba) on the Maccheronccini pasta quality. *Eur. Food Res. Technol.* **2016**, *242*, 477–485. [CrossRef]

33. Bilgicli, N. Some chemical and sensory properties of gluten-free noodle prepared with different legume, pseudocereal and cereal flour blends. *J. Food Nutr. Res.* **2013**, *52*, 251–255.

34. Torres, A.; Frias, J.; Granito, M.; Guerra, M.; Vidal-Valverde, C. Chemical, biological and sensory evaluation of pasta products supplemented with α-galactoside-free lupin flours. *J. Sci.* **2007**, *87*, 74–81. [CrossRef]

35. Fredrikson, M.; Biot, P.; Alminger, M.; Carlsson, N.; Sandberg, A. Production process for high-quality pea-protein isolate with low content of oligosaccharides and phytate. *J. Agric. Food Chem.* **2001**, *49*, 1208–1212. [CrossRef] [PubMed]

36. Laleg, K.; Cassan, D.; Barron, C.; Prabhasankar, P.; Micard, V. Structural, culinary, nutritional and anti-nutritional properties of high protein, gluten free, 100% legume pasta. *PLoS ONE* **2016**, *11*. [CrossRef]

37. Campos-Vega, R.; Loarca-Piña, G.; Oomah, B.D. Minor components of pulses and their potential impact on human health. *Food Res. Int.* **2010**, *43*, 461–482. [CrossRef]

38. Sparvoli, F.; Bollini, R.; Cominelli, E. Nutritional value. In *Grain Legumes*; De Ron Antonio, M., Ed.; Springer: New York, NY, USA, 2015; pp. 291–326.

39. Martinez-Villaluenga, C.; Frias, J.; Vidal-Valverde, C. Alpha-galactosides: Antinutritional factors or functional ingredients? *Crit. Rev. Food Sci.* **2008**, *48*, 301–316. [CrossRef]

40. Campos-Vega, R.; Bassinello, P.; Cardoso-Santiago, R.A.; Oomah, B.D.; Grumezescu, A.M.; Holban, A.M. Chapter 20. In *Therapeutic, Probiotic, and Unconventional Foods*; Grumezescu, A.M., Holban, A.M., Eds.; Academic Press: Cambridge, MA, USA; pp. 367–386. Available online: https://www.sciencedirect.com/science/article/pii/B9780128146255000194 (accessed on 3 March 2020).

41. Giuberti, G.; Gallo, A. Reducing the glycaemic index and increasing the slowly digestible starch content in gluten-free cereal-based foods: A review. *Int. J. Food Sci. Technol.* **2018**, *53*, 50–60. [CrossRef]

42. Zhao, Y.H.; Manthey, F.A.; Chang, S.K.; Hou, H.J.; Yuan, S.H. Quality characteristics of spaghetti as affected by green and yellow pea, lentil, and chickpea flours. *J. Food Sci.* **2005**, *70*, s371–s376. [CrossRef]

43. Frias, J.; Kovács, E.; Sotomayor, C.; Hedley, C.; Vidal-Valverde, C. Processing peas for producing macaroni. *LWT Food Sci. Technol.* **1997**, *204*, 66–71. [CrossRef]

44. Carcea, M.; Narducci, V.; Turfani, V.; Giannini, V. Polyphenols in raw and cooked cereals/pseudocereals/legume pasta and couscous. *Foods* **2017**, *6*, 80. [CrossRef]

45. Gull, A.; Prasad, K.; Kumar, P. Nutritional, antioxidant, microstructural and pasting properties of functional pasta. *J. Saudi Soc. Agric. Sci.* **2018**, *17*, 147–153. [CrossRef]

46. Verardo, V.; Arráez-Román, D.; Segura-Carretero, A.; Marconi, E.; Fernández-Gutiérrez, A.; Caboni, M.F. Determination of free and bound phenolic compounds in buckwheat spaghetti by RP-HPLC-ESI-TOF-MS: Effect of thermal processing from farm to fork. *J. Agric. Food Chem.* **2011**, *59*, 7700–7707. [CrossRef] [PubMed]

47. Rocchetti, G.; Lucini, L.; Chiodelli, G.; Giuberti, G.; Montesano, D.; Masoero, F.; Trevisan, M. Impact of boiling on free and bound phenolic profile and antioxidant activity of commercial gluten-free pasta. *Food Res. Int.* **2017**, *100*, 69–77. [CrossRef] [PubMed]

48. Del Bo, C.; Bernardi, S.; Marino, M.; Porrini, M.; Tucci, M.; Guglielmetti, S.; Cherubini, A.; Carrieri, B.; Kirkup, B.; Kroon, P.; et al. Systematic review on polyphenol intake and health outcomes: Is there sufficient evidence to define a health-promoting polyphenol-rich dietary pattern? *Nutrients* **2019**, *11*, 1355. [CrossRef]

49. Tresserra-Rimbau, A.; Rimm, E.B.; Medina-Remón, A.; Martínez-González, M.A.; de la Torre, R.; Corella, D.; Salas-Salvadó, J.; Gómez-Gracia, E.; Lapetra, J.; Arós, F.; et al. Inverse association between habitual polyphenol intake and incidence of cardiovascular events in the PREDIMED study. *Nutr. Metab. Cardiovasc. Dis.* **2014**, *24*, 639–647. [CrossRef]

50. Rosa-Sibakov, N.; Heiniö, R.-L.; Cassan, D.; Holopainen-Mantila, U.; Micard, V.; Lantto, R.; Sozer, N. Effect of bioprocessing and fractionation on the structural, textural and sensory properties of gluten-free faba bean pasta. *LWT Food Sci. Technol.* **2016**, *67*, 27–36. [CrossRef]

51. Bouasla, A.; Wójtowicz, A.; Zidoune, M.N.; Olech, M.; Nowak, R.; Mitrus, M.; Oniszczuk, A. Gluten-free precooked rice-yellow pea pasta: Effect of extrusion-cooking conditions on phenolic acids composition, selected properties and microstructure. *J. Food Sci.* **2016**, *81*, C1070–C1079. [CrossRef]

52. Jayasena, V.; Nasar-Abbas, S.M. Development and quality evaluation of high-protein and high-dietary-fiber pasta using lupin flour. *J. Texture Stud.* **2012**, *43*, 153–163. [CrossRef]

53. Bouasla, A.; Wójtowicz, A.; Zidoune, M.N. Gluten-free precooked rice pasta enriched with legumes flours: Physical properties, texture, sensory attributes and microstructure. *LWT* **2017**, *75*, 569–577. [CrossRef]

54. Sobota, A.; Zarzycki, P. Effect of pasta cooking time on the content and fractional composition of dietary fiber. *J. Food Qual.* **2013**, *36*, 127–132. [CrossRef]

55. Taherian, A.R.; Mondor, M.; Labranche, J.; Drolet, H.; Ippersiel, D.; Lamarche, F. Comparative study of functional properties of commercial and membrane processed yellow pea protein isolates. *Food Res. Int.* **2011**, *44*, 2505–2514. [CrossRef]

56. Tsai, P.J.; Sun, Y.F.; Hsiao, S.M. Strengthening the texture of dried guava slice by infiltration of phenolic compounds. *Food Res. Int.* **2010**, *43*, 825–830. [CrossRef]

Article

Effect of Grape Pomace Addition on the Technological, Sensory, and Nutritional Properties of Durum Wheat Pasta

Roberta Tolve [1], Gabriella Pasini [2], Fabiola Vignale [1], Fabio Favati [1] and Barbara Simonato [1,*]

1   Department of Biotechnology, University of Verona, Strada Le Grazie 15, 37134 Verona, Italy;
    roberta.tolve@univr.it (R.T.); fabiola.vignale@gmail.com (F.V.); fabio.favati@univr.it (F.F.)
2   Department of Agronomy, Food, Natural Resources, Animals and Environment, University of Padova,
    Viale dell'Università 16, 35020 Legnaro Padova, Italy; gabriella.pasini@unipd.it
*   Correspondence: barbara.simonato@univr.it

check for updates

Received: 21 February 2020; Accepted: 17 March 2020; Published: 19 March 2020

Abstract: In this study, fortified pasta was prepared by replacing semolina with 0, 5, and 10 g/100 g of grape pomace (GP), a food industry by-product, rich in fiber and phenols. GP inclusion in pasta significantly reduced its optimum cooking time and the swelling index, while also increasing the cooking loss ($p < 0.05$). Furthermore, pasta firmness and adhesiveness were enhanced by the GP addition, as well as the total phenol content and the antioxidant activity, evaluated through ABTS and FRAP assays ($p < 0.05$). From a nutritional point of view, increasing amounts of GP resulted in a significant decrease in the rapidly digestible starch and an increase in the slowly digestible starch, while the predicted in vitro glycemic index was also reduced ($p < 0.05$). Sensory analysis showed that fortified spaghetti had good overall acceptability, and the results suggest that GP-fortified pasta could represent a healthy product with good technological and sensory properties.

Keywords: agro-industrial by-product; fortified pasta; dietary fiber; phenolic compounds; starch digestibility

1. Introduction

Pasta, a staple food consumed worldwide, can represent an excellent choice for the addition of bioactive compounds [1,2]. Considering the concepts associated with the circular economy, the possibility of using food industry by-products as a source of bioactive compounds, such as antioxidants and dietary fibers [3,4], is of interest. Among the various food industry by-products with potential healthy properties, grape pomace (GP), a residue of grape processing in wine production, may be an interesting ingredient. GP represents about 20% of the mass of total processed grapes and it is estimated that for every 100 liters of produced wine, about 17 kg of GP must be disposed. Grape skins represent more than 80% of the wet weight of GP and being rich in phenolic compounds and dietary fiber may be an interesting and cheap source of healthy moieties [5–7]. In humans, phenolic compounds exert a wide range of beneficial physiological activities, and many epidemiological studies associate a phenolic-rich diet with the prevention of several pathologies, such as cardiovascular diseases, diabetes, as well as some types of cancer [8–10]. GP also contains a noticeable quantity of dietary fiber (DF), whose level depends on several factors, among which is the grape variety. For red grapes, the DF content has been reported to range between 51% and 74% (by weight on dry matter) and the DF consumption may help in reducing the incidence of some types of cancer, as well as the development of diabetes. Moreover, fibers improve satiety and intestinal peristalsis, favor blood cholesterol decrease, and prevent obesity [11–14]. While the recommended DF intake is 25–30 g per day, this value is often not reached because of the modern eating habits, and the availability of fiber-enriched or fortified foods may represent a good opportunity for consumers to increase their daily fiber intake [15].

Nowadays, pasta is a product consumed worldwide and represents a remarkable staple food to convey bioactive compounds. To achieve this, several studies have been carried out to develop pasta with enhanced nutritional properties by using, for instance, olive pomace, carrot pomace, and tomato. The addition of these ingredients in pasta formulation, as well as the increase in dietary fiber and antioxidant activity, generally result in the modification of the technological and cooking properties, starch digestibility, and glycemic index [4,16–18]. In this framework, GP can represent a valuable ingredient to produce fortified food items, and this study aimed to evaluate the effects of replacing durum wheat semolina with different levels of GP in the production of durum wheat spaghetti (0/100, 5/95, and 10/90 g GP/g semolina). The effects on pasta quality were assessed for cooking properties, color, and texture. The nutritional properties of GP-fortified pasta were also evaluated, considering the total polyphenol content, the antioxidant capacity and the "in vitro" starch digestibility. Finally, the sensory properties of the fortified prepared spaghetti were also appraised.

2. Materials and Methods

2.1. Grape Pomace Powder Preparation and Chemical Composition

Grape pomace from *Vitis vinfera* L cv. Corvina, a red grape used for Amarone wine production, was kindly supplied by Tinazzi srl (Verona, Italy). The GP, after alcoholic fermentation, was separated, pressed, and immediately recovered and dried in a vacuum oven (VD 115 Binder GmbH, Tuttlingen, Germany) (40 °C, 30 kPa) until reaching a final moisture content of 11.0 g water/100 g GP. Afterwards, the stems and seeds were manually removed from the pomace and the by-product was milled (GM200 Retsch, Haan, Germany) to obtain a powder with a particle size < 0.2 mm. The GP was then stored in an airtight, dark plastic container until analyzed or used for pasta preparation. The GP chemical composition was assessed in triplicate according to the following AOAC standard methods: 930.15 for moisture, 976.05 for protein, 985.29 for total fiber, and 942.05 for ash content [19]. All chemical components were expressed as g/100 g of dry matter (DM).

2.2. Pasta Preparation

Commercial durum wheat semolina was purchased in a local market, and the following is the nutrient composition reported on the label: carbohydrates 71.8 g/100 g, protein 11 g/100 g, fat 1.8 g/100 g, fiber 3 g/100 g. Pasta was prepared, replacing semolina with different GP amounts, obtaining the samples GP0, GP5, and GP10 (0/100, 5/95 and 10/90 g GP/g semolina). The dough was prepared using a professional pasta machine (Mod. Lillodue, Bottene, Marano Vicentino, Italy) by adding 35% of tap water at 40 °C to pure semolina or to the blend GP-semolina. The dough was mixed for 10 min before being extruded through a 1.75-mm bronze spaghetti die, cutting the spaghetti at the standard industrial length of 250 mm. The spaghetti were air-dried at 50 °C in a Juan laboratory drier (Thermo Fisher Scientific, Waltham, MA, USA) until reaching a residual moisture content of 12 g water/100 g pasta. The pasta samples were then cooked in boiling distilled water in a ratio 1:10 (w/v). The optimum cooking time was assessed experimentally.

2.3. Pasta Properties Determination

Pasta moisture content, optimum cooking time (OCT), and cooking loss (CL) were assessed according to the AACC methods 44-15A and 66-50 [20]. The swelling index (SI) was measured according to the procedure reported by Clearly and Brennan [21].

2.4. Color Analysis

Uncooked and cooked spaghetti color was measured with a reflectance colorimeter (Minolta Chroma meter CR-300, Osaka, Japan) (illuminant D65) following the CIE - L* a* b* color system. Each measure was made in triplicate.

2.5. Texture Analysis

A TA-XT plus Texture Analyser (Stable Micro Systems, Godalming, UK) equipped with a 5-kg load cell was used to measure the cooked pasta firmness and adhesiveness. The firmness test was performed according to the AACC 16-50 method [20]. Samples were put side by side on the lower plate perpendicularly to a probe and compressed by a superior plate (speed 2 mm/s, percentage of deformation 75%). The firmness was recorded as the maximum force required to compress the pasta samples, and the adhesiveness was defined as the negative peak force required to separate the probe from the sample surface.

2.6. Determination of Total Phenolic Compounds, ABTS, and FRAP Assay

One gram of each sample (GP, GP0, GP5, GP10) was extracted for 24 h, at room temperature, with 15 mL of MeOH:HCl (97:3) under continuous stirring in the dark [22]. After centrifugation at 3500 g for 10 min at 4 °C, the supernatant was recovered and utilized for assessing the total phenolic content (TPC), ABTS and FRAP radical scavenging activities. The TPC was determined according to Singleton and Rossi [23] with slight modifications. In detail, 0.2 mL of the extract was mixed with 0.2 mL of Folin-Ciocalteau reagent. After 5 min, 4 mL of Na_2CO_3 solution (0.7 M) was added and the final volume was brought to 10 mL using Milli-Q water. After 1 h, the absorbance of the solution was measured at 725 nm (ATi Unicam UV2, Akribis Scientific, Cambridge, UK). The TPC was expressed as mg of gallic acid equivalent (GAE)/g of dry matter (DM). The ABTS assay was performed according to the method proposed by Del Pino-García et al. [22]. A stock solution of the radical cation ($ABTS^{\bullet+}$) was prepared by incubating in the dark for 12 h, at room temperature, with ABTS (7 mM) and $K_2S_2O_8$ (2.45 mM) (1:1 ratio). Subsequently, 0.2 mL of the methanolic extract obtained as described were added to 9.8 mL of $ABTS^{\bullet+}$ working solution and incubated at room temperature, in the dark, for 30 min. Absorbance was measured spectrophotometrically at 734 nm. The results were expressed as the μM of Trolox equivalent (TE)/g of DM, using a Trolox calibration curve. The FRAP assay was performed according to Benzie and Strain [24]. The FRAP reagent was obtained by mixing in a volume ratio of 10:1:1, a sodium acetate buffer (300 mM, pH 3.6), TPTZ solution (10 mM) in HCl (40 mM), and a $FeCl_3.6H_2O$ solution (20 mM). A total of 10 μL of the methanolic extract were mixed with 1 mL of MilliQ water and 1.8 mL of FRAP reagent. The absorbance was then measured at 593 nm and the results expressed as μM of TE/g of DM, using a Trolox calibration curve.

2.7. Starch Fractions Determination

Rapidly digestible starch (RDS) and slowly digestible starch (SDS) were measured in cooked pasta samples according to the method of Englyst et al. [25], slightly modified. Briefly, the starch hydrolysis of pasta cooked to the optimum was performed at 37 °C using an enzyme mixture composed by pancreatic amylase (1350 U), amyloglucosydase (3300 U), and invertase (2000 U). The amount of glucose released after pasta starch hydrolysis was measured spectrophotometrically at 510 nm using a glucose oxidase kit (GOPOD, Megazyme, Ireland). The RDS value was calculated considering the amount of glucose released after 20 min of enzymatic reaction. SDS was determined by subtracting the glucose released after 20 min from the glucose released after 120 min. The resistant starch (RS) was measured according to the Megazyme protocol (K-RSTAR, Megazyme, Ireland).

2.8. Hydrolysis Index and Predicted Glycemic Index

The hydrolysis index was determined according to Simonato et al. [4]. Briefly, 100 mg of cooked pasta was incubated at 37 °C in glass vials, adding 4 mL of maleic buffer (pH 6) containing 40 mg of pancreatic α-amylase (3000 U/mg) and 4 μL of amyloglucosidase solution (300 U/mL) (Megazyme Ltd.). After 0, 30, 60, 120, and 180 min, the reaction was stopped by adding 4 mL of pure ethanol, and the solution was centrifuged at 2500 g for 10 min. D-Glucose was assessed as described above. The hydrolysis index (HI) was considered as the percentage between the area under the hydrolyses curve (0–180 min) of each pasta sample and the corresponding area of a white bread curve, used as a reference. The predicted glycemic index (pGI) was measured according to the formula proposed by Granfeldt et al. [26]: $pGI = 8.198 + 0.862 \times HI$.

2.9. Scanning Electron Microscopy

The microstructure of cross sections of uncooked and cooked pasta samples were observed using a TM-1000 Environmental Electronic Scanning Microscope (Hitachi High-Technology, Tokyo, Japan) equipped with an accessory door tilt and rotate port plate (Deben, UK) at room temperature. The pasta samples were viewed using 1200× magnification, and the images were processed with AdobePhotoshop 6.0.

2.10. Sensory Evaluation

The sensory profile of pasta samples was assessed by using a panel of 30 individuals (14 men, 16 women; 23–54 years old) and the informed consent was obtained from each subject prior to their participation in the study. The judges were trained to recognize different intensities of 12 sensory attributes (Appearance: typical semolina pasta color, color uniformity, starch release; Taste: sweet, acid; Aroma: pasta, wine, acid; Texture: astringent, roughness, adhesiveness, graininess). Each judge received 15 g of cooked pasta placed in a container with a lid. The order of sample presentation was balanced and randomized among judges. A 9-point scale, in which 1 represented the lowest intensity and 9 the highest intensity for each attribute, was used. Mean sensory scores from the 30 panel members for each attribute were then calculated. Panelists were asked also to give a comment on the overall acceptability of pasta.

2.11. Statistical Analysis

The analyses were carried out in triplicates and the mean values ± standard deviation are reported. The variables were tested for significance using one-way analysis of variance (ANOVA). Differences among means were assessed using Tukey's HSD test ($p < 0.05$). All the statistical analyses were carried out using the statistical software R project (version 3.2.3 December 2015) [27].

3. Results and Discussion

3.1. Grape Pomace Composition

The assessed chemical composition of the dried GP was as follows: moisture (11.0 ± 0.2 g/100 g DM), protein (11.19 ± 0.97 g/100 g DM), total fiber (52.3 ± 2.1 g/100 g DM), ash (4.17 ± 0.87 g/100 g DM).

3.2. Cooking, Textural Properties, and Color Value

Cooking and textural properties of control and fortified pasta samples are reported in Table 1. GP addition caused a significant reduction in the OCT values ($p < 0.05$), and this could be explained considering that high GP levels may cause a decrease in the overall gluten quantity, affecting the starch–protein structure and, hence, the texture and cooking properties of the spaghetti [28]. The high GP fiber content (52.3 ± 2.1 g/100 g GP), in comparison to the content of the control (3 g/100 g) could contribute to altering the gluten matrix, therefore allowing swift water entry into the pasta central core during cooking and inducing an earlier starch gelatinization and OCT reduction. CL is an index of the capability of the pasta starch–protein matrix to retain its structural integrity during cooking [29].

Table 1. Cooking quality parameters and texture analysis of control pasta (GP0) and pasta fortified with different percentages of grape pomace (GP5 and GP10). The values are reported as mean ± standard deviation.

Pasta Samples	Optimum Cooking Time (min)	Cooking Loss (%)	Swelling Index (g Water/g Dry Pasta)	Firmness (N)	Adhesiveness (N)
GP0	6.0	6.61 ± 0.03 [a]	3.59 ± 0.08 [a]	104.20 ± 0.01 [a]	−0.11 ± 0.01 [a]
GP5	5.5	8.18 ± 0.23 [b]	2.98 ± 0.21 [b]	113.44 ± 0.02 [b]	−0.45 ± 0.02 [b]
GP10	5.0	9.48 ± 0.10 [c]	0.97 ± 0.03 [c]	135.80 ± 0.02 [c]	−0.66 ± 0.02 [c]

Values with different superscripts within the same column are significantly different for $p < 0.05$.

It has been reported that the addition of fiber can cause a CL increase as fiber may interfere with the starch gluten network, causing its weakening. Consequently, starch gelatinization could occur more rapidly, justifying the OCT reduction and higher leaching of the gelatinized starch from the pasta into the water during cooking [30,31]. The experimental data showed that GP addition caused a significant CL increase, both in GP5 and GP10 spaghetti ($p < 0.05$). Pasta fortification caused a significant reduction of the SI value ($p < 0.05$) that could be ascribed to the competition for water between fiber and starch during pasta cooking. Fiber addition through GP fortification also caused an increase in spaghetti firmness ($p < 0.05$). A similar trend has been reported in the literature when studying the enrichment of durum wheat spaghetti with Barley BalanceTM, even if opposite results have been reported when using different sources of fiber (e.g., inulin, guar gum) [28]. Adhesiveness also showed a significant increment in the GP5 and GP10 samples with respect to plain spaghetti. This could be ascribed to the breaking of the continuous structure of the pasta due to the fiber addition [32].

With regards to the pasta color, Table 2 shows that a progressive increase in GP concentration led to a significant reduction in lightness (L*) both in uncooked and cooked spaghetti. An increase in redness (a*) was also evident along with a GP increase for both uncooked and cooked pasta with a major extent for the latter. Moreover, redness values for uncooked GP5 and GP10 were not significantly different, unlike what was observed in the same samples of cooked pasta. A significant decrease in yellowness (b*) was evident with the progressive addition of GP in both cooked and uncooked spaghetti compared to GP0.

Table 2. Color analysis of uncooked and cooked control pasta (GP0) and pasta fortified with different percentages of grape pomace (GP5 and GP10). The data, reported as mean ± standard deviation, are expressed as L*, a*, and b* values.

Pasta Samples	L*		a*		b*	
	Uncooked	Cooked	Uncooked	Cooked	Uncooked	Cooked
GP0	65.95 ± 2.45 [a]	70.13 ± 1.90 [a]	−0.61 ± 0.34 [b]	−5.57 ± 0.18 [c]	16.18 ± 0.92 [a]	14.24 ± 1.19 [a]
GP5	47.64 ± 2.57 [b]	43.28 ± 0.52 [b]	1.46 ± 0.45 [a]	4.84 ± 0.32 [b]	6.50 ± 0.44 [b]	6.06 ± 0.64 [b]
GP10	43.55 ± 3.03 [b]	33.43 ± 1.75 [c]	1.62 ± 0.04 [a]	6.61 ± 0.60 [a]	5.27 ± 0.11 [c]	5.27 ± 0.10 [b]

Values with different superscripts within the same column are significantly different for $p < 0.05$.

3.3. Polyphenols and Antioxidant Activity

The TPC quantity of GP powder was 34.45 ± 0.38 mg GAE/g DM, while the antioxidant activity was 393.43 ± 2.58 and 171.18 ± 0.74 µM TE/g DM evaluated by FRAP and ABTS, respectively. In the fortified spaghetti, the TPC content, as well as the antioxidant activity assessed by ABTS and FRAP assays, showed to be significantly higher in comparison to the control ($p < 0.05$) (Table 3). The TPC and antioxidant activity were positively correlated to the quantity of GP added, both in uncooked and cooked to the optimum spaghetti (r > 0.87). The cooking treatment caused a significant decrease ($p < 0.05$) in the antioxidant activity and of the assessed TPC values, whose reduction ranged from

24 to about 30% in GP5 and GP10 samples. Similar results are reported in the literature and could be ascribed to the degradation of phenolic compounds during cooking or to their leaching into the cooking water [4,17].

Table 3. Total phenolic component (TPC) and antioxidant activity (FRAP and ABTS) of cooked and uncooked control pasta (GP0) and pasta fortified with different percentages of grape pomace (GP5 and GP10). The values are expressed as mean ± standard deviation.

Pasta Samples	TPC (mg GAE/g dw)	FRAP (μM TE/g dw)	ABTS (μM TE/g dw)
Uncooked GP0	0.43 ± 0.03 [e]	0.30 ± 0.09 [e]	2.85 ± 0.06 [c]
Uncooked GP5	1.38 ± 0.04 [c]	8.27 ± 0.27 [c]	3.63 ± 0.18 [b]
Uncooked GP10	2.57 ± 0.06 [a]	11.32 ± 0.20 [a]	4.50 ± 0.20 [a]
Cooked GP0	0.15 ± 0.02 [f]	0.09 ± 0.07 [f]	0.60 ± 0.04 [e]
Cooked GP5	1.05 ± 0.14 [d]	7.34 ± 0.32 [d]	2.46 ± 0.22 [d]
Cooked GP10	1.81 ± 0.11 [b]	9.42 ± 0.21 [b]	4.40 ± 0.15 [a]

Values with different superscripts within the same column are significantly different for $p < 0.05$.

However, despite the recorded TPC losses, cooked pasta still has a good antioxidant capacity due to the polyphenol compounds retained in the structure. GP fortification caused a significant increase in the antioxidant activity of the pasta samples, both in the uncooked and cooked states ($p < 0.05$), and the measured antioxidant activity showed to be strictly correlated to the quantity of GP added to the dough according to both ABTS ($r = 0.82$) or FRAP ($r = 0.93$) assay (Table 3).

3.4. In Vitro Starch Digestibility

The evaluation of the nutritional quality of GP-fortified pasta, in terms of starch digestibility, was carried out using a well-established in vitro assay, validated also in vivo and recognized by the European Food Safety Authority [33]. On the base of the hydrolysis rate and extent, starch can be classified into different fractions: rapidly digestible starch (RDS), slowly digestible starch (SDS), and resistant starch (RS) [25]. The obtained data are reported in Table 4, together with the predicted glycemic index. In comparison to the control, increasing GP amounts in the dough caused a reduction in the RDS value, responsible for a rapid increment of glucose and insulin levels in the blood. The reduction was statistically significant for the GP10 sample, the RDS value being about 13% lower than that assessed in the control ($p < 0.05$). SDS is the starch fraction that undergoes slow digestion, allowing to maintain the glucose level in the blood over time and being a good indicator of the glycemic response in humans [34].

Table 4. Starch fractions expressed as percentages of the total starch and predicted glycemic index (pGI) of cooked control pasta (GP0) and cooked pasta fortified with different percentages of grape pomace (GP5 and GP10). RDS: rapidly digested starch; SDS: slowly digested starch; RS: resistant starch. The values are reported as mean ± standard deviation.

Cooked Pasta Samples	Starch Fractions (%)			pGI
	RDS	SDS	RS	
GP0	33.45 ± 1.16 [a]	32.94 ± 0.99 [a]	2.13 ± 0.16 [a]	57.46 ± 0.41 [a]
GP5	31.87 ± 1.64 [ab]	33.33 ± 1.36 [a]	2.19 ± 0.17 [a]	55.56 ± 0.22 [b]
GP10	29.09 ± 2.05 [b]	36.19 ± 0.34 [b]	1.83 ± 0.41 [a]	53.15 ± 1.37 [c]

Values with different superscripts within the same column are significantly different for $p < 0.05$.

The SDS level was also affected by increasing the amount of GP in spaghetti, reaching a significant increment of about 10% in the GP10 pasta sample ($p < 0.05$). With regards to the RS, the experimental data showed a reduction trend with the use of increasing GP amount in pasta. However, no significant differences could be highlighted among the various samples. The pGI, useful for predicting the likely in vivo glycemic response, showed a significant decrease ($p < 0.05$) with the progressive addition of

GP in pasta. Overall the data reported in Table 4 indicate that GP inclusion in pasta positively affected the rate of starch digestibility. The lower glucose release observed in this study could be partially due to the reduction in the starch content in the pasta because of the replacement of semolina with different GP amounts or the starch leaching into the water during cooking, as also reported in the literature [4,30]. It should be pointed out that GP comprises about 52% of fiber that can compete with the starch granules for water adsorption. Therefore, starch gelatinization could be reduced, as well as the action of the starch hydrolyzing enzymes, resulting in a limited starch digestibility [35]. Moreover, GP was rich also in phenolic compounds, such as phenolic acids, flavonoids, and tannins, that can inhibit the α-amylase and α-glucosidase activities, as previously reported by Hanhineva et al. [36]. Hence, the TPC increase in GP pasta could also contribute to the decrease in the rate of starch digestion.

3.5. Scanning Electron Microscopy (ESEM)

ESEM observations of cross sections of control (GP0), GP5, and GP10 spaghetti reveal a well-developed protein matrix with circular and lentil-shaped starch granules trapped in the gluten network (Figure 1).

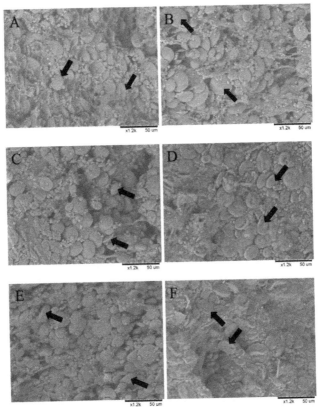

Figure 1. SEM micrographs of uncooked (left) and cooked (right) GP0 (**A,B**), GP5 (**C,D**), and GP10 (**E,F**) pasta samples. The black arrows indicate starch granules.

The structural differences observed in raw pasta are more evident after cooking, especially for the GP10 spaghetti sample, where the protein network appears to be less structured. GP0, GP5, and GP10 spaghetti show a similar pattern, in which swollen starch granules are completely embedded in the matrix. However, the images of the cooked fortified samples highlight a lower presence of a well-structured filamentous protein network in comparison with the control.

3.6. Sensory Evaluation

Sensory evaluation of the spaghetti samples revealed that the substitution of semolina wheat with GP powder significantly affected most of the selected attributes (Figure 2).

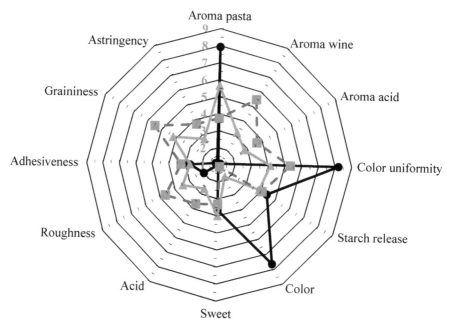

Figure 2. Sensory scores of quality attributes of pasta fortified with grape pomace at different addition levels (GP0 black solid line; GP5 solid gray line; GP10 dashed gray line).

Using increasing amounts of GP, the perception of the aroma pasta and semolina pasta color decreased ($p < 0.05$). Aroma wine, aroma acid, flavor acid, astringency, and graininess, not or slightly detected in the control sample, were significantly perceived, with the GP10 sample being significantly different from the GP5 one. Roughness was also influenced by the GP addition, and the fortified samples were perceived as being rougher than the control ($p < 0.05$). As demonstrated by the instrumental evaluation of pasta color, the inclusion of GP caused a significant variation that the panelists perceived regardless of the GP concentration used. As far as color uniformity, GP addition caused its reduction ($p < 0.05$) and the lowest score was recorded for the GP5 sample, where the amount of added GP could not uniformly hinder the native semolina color. For starch release, adhesiveness, and sweet the use of GP did not cause any statistical difference. However, a slight adhesiveness increase in GP-fortified spaghetti was assessed by the panel, even if not significant. This is somehow in contrast with the adhesiveness values measured using the Texture Analyzer, which highlighted an adhesiveness increase with GP addition in the dough (Table 1). Similar results have been reported by Shogren et al. [37] when evaluating the firmness perception in spaghetti fortified with soy flour. Moreover, no matter the GP amounts added, the panelist judged as acceptable the overall quality of the fortified pasta.

4. Conclusions

The results of this study showed that pasta fortification with GP caused an increase in the TPC and antioxidant activity in the cooked product. GP addition also affected the rate of starch digestibility, with a decrement of the RDS and an increment of the SDS, while the pGI, useful for predicting the likely in vivo glycemic response, showed a significant decrease with increasing amounts of GP in spaghetti. GP addition also influenced the pasta technological properties, increasing cooking loss, firmness, and adhesiveness in the cooked samples, while a decrease of the swelling index was observed. Furthermore, while pasta produced using only semolina had a fiber content of about 3 g fiber/100 g pasta, in the GP5- and GP10-fortified spaghetti, the fiber level ranged from 5.6 to 8.2 g fiber/100 g pasta, respectively. This is of the utmost interest because, according to the European Union, legislating a foodstuff as a *"source of fiber"* or *"high fiber"* is only possible if it contains at least 3 or 6 g of dietary fiber/100 g of product, respectively [38]. Moreover, the newly formulated pasta also showed good sensory acceptability. In conclusion, GP, a winemaking by-product available at low cost in great amounts, may represent an interesting ingredient to produce a functional pasta rich in antioxidant compounds and dietary fiber, with a potentially positive impact on the human health, also due to the assessed lower pGI in comparison with pure semolina spaghetti. While this study was focused on the use of GP obtained from a red grape variety rich in phenols, it would be of interest to carry out further studies to verify if the use of GP of different grape varieties may give the same positive results. Additionally, future studies should aim to assess the phenols bioaccessibility through in vitro digestion, as well as to characterize the different phenols moieties by using LC-MS analysis. The interaction between grape pomace and intestinal bacteria, as well as the healthy consequences, should be also investigated.

Author Contributions: Conceptualization, B.S.; Formal analysis, B.S., G.P., R.T. and F.V.; Methodology B.S., G.P., R.T. and F.V.; Data Curation, B.S. and R.T.; Writing–Original Draft Preparation, B.S. and T.R.; Writing–Review & Editing, F.F. All authors have read and agreed to the published version of the manuscript.

Funding: This research received no external funding.

Conflicts of Interest: The authors declare no conflict of interest.

References

1. Oliviero, T.; Fogliano, V. Food design strategies to increase vegetable intake: The case of vegetable enriched pasta. *Trends Food Sci. Technol.* **2016**, *51*, 58–64. [CrossRef]
2. Spinelli, S.; Padalino, L.; Costa, C.; Del Nobile, M.A.; Conte, A. Food by-products to fortified pasta: A new approach for optimization. *J. Clean. Prod.* **2019**, *215*, 985–991. [CrossRef]
3. Padalino, L.; D'Antuono, I.; Durante, M.; Conte, A.; Cardinali, A.; Linsalata, V.; Mita, G.; Logrieco, A.F.; Del Nobile, M.A. Use of Olive Oil Industrial By-Product for Pasta Enrichment. *Antioxidants* **2018**, *7*, 59. [CrossRef]
4. Simonato, B.; Trevisan, S.; Favati, F.; Tolve, R.; Pasini, G. Pasta fortification with olive pomace: Effects on the technological characteristics and nutritional properties. *Food Sci. Technol.* **2019**, *114*, 108368. [CrossRef]
5. Sant'Anna, V.; Christiano, F.D.P.; Marczak, L.D.F.; Tessaro, I.C.; Thys, R.C.S. The effect of the incorporation of grape marc powder in fettuccini pasta properties. *Food Sci. Technol.* **2014**, *58*, 497–501. [CrossRef]
6. Zanotti, I.; Dall'Asta, M.; Mena, P.; Mele, L.; Bruni, R.; Ray, S.; Del Rio, D. Atheroprotective effects of (poly)phenols: A focus on cell cholesterol metabolism. *Food Funct.* **2015**, *6*, 13–31. [CrossRef]
7. Hogervorst, J.C.; Miljić, U.; Puškaš, V. Extraction of bioactive compounds from grape processing by-products. In *Handbook of Grape Processing by-Products*; Galanakis, C.M., Ed.; Academic Press: Amsterdam, NL, USA, 2017; pp. 105–135.
8. Montealegre, R.R.; Peces, R.R.; Chacòn Vozmediano, J.L.; Martìnez Gascuena, J.; Garcìa Romero, E. Phenolic compounds in skins and seeds of ten grape *Vitis vinifera* varieties grown in a warm climate. *J. Food Comp. Anal.* **2006**, *19*, 687–693. [CrossRef]
9. Cao, H.; Ou, J.; Chen, L.; Zhang, Y.; Szkudelski, T.; Delmas, D.; Daglia, M.; Xiao, J. Dietary polyphenols and type 2 diabetes: Human study and clinical trial. *Crit. Rev. Food Sci. Nutr.* **2019**, *59*, 3371–3379. [CrossRef]

10. Rothwell, J.A.; Knaze, V.; Zamora-Ros, R. Polyphenols: Dietary assessment and role in the prevention of cancers. *Curr. Opin. Clin. Nutr. Metab. Care* **2017**, *20*, 512–521. [CrossRef]

11. Zhang, H.; Wang, H.; Cao, X.; Wang, J. Preparation and modification of high dietary fiber flour: A review. *Food Res. Int.* **2018**, *113*, 24–35. [CrossRef]

12. Llobera, A.; Cañellas, J. Dietary fibre content and antioxidant activity of Manto Negro red grape (*Vitis vinifera*): Pomace and stem. *Food Chem.* **2007**, *101*, 659–666. [CrossRef]

13. Deng, Q.; Penner, M.H.; Zhao, Y.Y. Chemical composition of dietary fiber and polyphenols of five different varieties of wine grape pomace skins. *Food Res. Int.* **2011**, *44*, 2712–2720. [CrossRef]

14. Tseng, A.; Zhao, Y. Wine grape pomace as antioxidant dietary fibre for enhancing nutritional value and improving storability of yogurt and salad dressing. *Food Chem.* **2013**, *138*, 356–365. [CrossRef] [PubMed]

15. Jane, M.; McKay, J.; Pal, S. Effects of daily consumption of psyllium, oat bran and polyGlycopleX on obesity-related disease risk factors: A critical review. *Nutrition* **2019**, *57*, 84–91. [CrossRef] [PubMed]

16. Gull, A.; Prasad, K.; Kumar, P. Effect of Millet Flours and Carrot Pomace on Cooking Qualities, Color and Texture of Developed Pasta. *LWT Food Sci. Technol.* **2015**, *63*, 470–474. [CrossRef]

17. Gull, A.; Prasad, K.; Kumar, P. Nutritional, antioxidant, microstructural and pasting properties of functional pasta. *J. Saudi Soc. Agric. Sci.* **2018**, *17*, 147–153. [CrossRef]

18. Pasqualone, A.; Gambacorta, G.; Summo, C.; Caponio, F.; Di Miceli, G.; Flagella, Z.; Marrese, P.P.; Piro, G.; Perrotta, C.; De Bellis, L.; et al. Functional, textural and sensory properties of dry pasta supplemented with lyophilized tomato matrix or with durum wheat bran extracts produced by supercritical carbon dioxide or ultrasound. *Food Chem.* **2016**, *213*, 545–553. [CrossRef]

19. AOAC. *Official Methods of Analysis*, 17th ed.; AOAC International: Rockville, MD, USA, 2000.

20. AACC. *Approved Methods of the AACC*, 10th ed.; American Association of Cereal Chemists: St. Paul, MN, USA, 2000.

21. Cleary, L.; Brennan, C. The influence of a $(1{\to}3)(1{\to}4)$-β-d-glucan rich fraction from barley on the physico-chemical properties and in vitro reducing sugars release of durum wheat pasta. *Int. J. Food Sci. Technol.* **2006**, *41*, 910–918. [CrossRef]

22. Del Pino-García, R.; González-Sanjosé, M.L.; Rivero-Pérez, M.D.; García-Lomillo, J.; Muñiz, P. The effects of heat treatment on the phenolic composition and antioxidant capacity of red wine pomace seasonings. *Food Chem.* **2017**, *221*, 1723–1732. [CrossRef]

23. Singleton, V.L.; Rossi, J.A. Colorimetry of total phenolics with phosphomolybdic-phosphotungstic acid reagents. *Am. J. Enol. Vitic.* **1965**, *16*, 144–158.

24. Benzie, I.F.; Strain, J.J. The ferric reducing ability of plasma (FRAP) as a measure of "antioxidant power": The FRAP assay. *Anal. Biochem.* **1996**, *239*, 70–76. [CrossRef] [PubMed]

25. Englyst, H.N.; Kingman, S.M.; Cummings, J.H. Classification and measurement of nutritionally important starch fractions. *Eur. J. Clin. Nutr.* **1992**, *46*, S33–S50. [PubMed]

26. Granfeldt, Y.; Bjorck, I.; Drews, A.; Tovar, J. An in vitro procedure based on chewing to predict metabolic response to starch in cereal and legume products. *Eur. J. Clin. Nutr.* **1992**, *46*, 649–660. [CrossRef] [PubMed]

27. R Core Team. *R: A Language and Environment for Statistical Computing*; R Foundation for Statistical Computing: Vienna, Austria, 2015.

28. Rakhesh, N.; Fellows, C.M.; Sissons, M. Evaluation of the technological and sensory properties of durum wheat spaghetti enriched with different dietary fibers. *J. Sci. Food Agric.* **2015**, *95*, 2–11. [CrossRef] [PubMed]

29. Song, X.; Zhu, W.; Pei, Y.; Ai, Z.; Chen, J. Effects of wheat bran with different colors on the qualities of dry noodles. *J. Cereal Sci.* **2013**, *58*, 400–407. [CrossRef]

30. Aravind, N.; Sissons, M.; Egan, N.; Fellows, C. Effect of insoluble dietary fiber addition on technological, sensory, and structural properties of durum wheat spaghetti. *Food Chem.* **2012**, *130*, 299–309. [CrossRef]

31. Kim, S.H.; Lee, J.W.; Heo, Y.; Moon, B.K. Effect of Pleurotus eryngii mushroom-glucan on quality characteristics of common wheat pasta. *J. Food Sci.* **2016**, *81*, 835–840. [CrossRef]

32. Kim, B.; Kim, S.; Bae, G.; Chang, M.B.; Moon, B. Quality characteristics of common wheat fresh noodle with insoluble dietary fiber from kimchi by-product. *Food Sci. Technol.* **2017**, *85*, 240–245. [CrossRef]

33. EFSA. Scientific Opinion on the substantiation of a health claim related to "Slowly digestible starch in starch-containing foods" and "reduction of post-prandial glycemic responses". *EFSA J.* **2011**, *9*, 2292.

34. Zhang, G.; Hamaker, B.R. Slowly digestible starch: Concept, mechanism, and proposed extended glycemic index. *Crit. Rev. Food Sci. Nutr.* **2009**, *49*, 852–867. [CrossRef]

35. Kim, E.H.J.; Petrie, J.R.; Motoi, L.; Morgenstern, M.P.; Sutton, K.H.; Mishra, S.; Simmons, L.D. Effect of structural and physicochemical characteristics of the protein matrix in pasta on in vitro starch digestibility. *Food Biophys.* **2008**, *3*, 229–234. [CrossRef]

36. Hanhineva, K.; Törrönen, R.; Bondia-Pons, I.; Pekkinen, J.; Kolehmainen, M.; Mykkänen, H.; Poutanen, K. Impact of Dietary Polyphenols on Carbohydrate Metabolism. *Int. J. Mol. Sci.* **2010**, *11*, 1365–1402. [CrossRef] [PubMed]

37. Shogren, R.L.; Hareland, G.A.; Wu, Y.V. Sensory evaluation and composition of spaghetti fortified with soy flour. *J. Food Sci.* **2006**, *71*, S428–S432. [CrossRef]

38. Commission Regulation. (EC) No 1924/2006 of the European Parliament and of the Council of 20 December 2006 on nutrition and health claims made on foods. *Off. J. Eur. Union L* **2006**, *404*, 9–25.

MDPI

St. Alban-Anlage 66

4052 Basel

Switzerland

Tel. +41 61 683 77 34

Fax +41 61 302 89 18

www.mdpi.com

Foods Editorial Office

E-mail: foods@mdpi.com

www.mdpi.com/journal/foods

Lightning Source UK Ltd.
Milton Keynes UK
UKHW050149270821
389529UK00003B/145